Field Guide to Freshwater Invertebrates
of North America

FIELD GUIDE TO FRESHWATER INVERTEBRATES OF NORTH AMERICA

James H. Thorp
Kansas Biological Survey and
Department of Ecology and Evolutionary Biology
University of Kansas, Lawrence, KS, USA

D. Christopher Rogers
Kansas Biological Survey, University of Kansas,
and Invertebrate Zoology Services, Lawrence, KS, USA

AMSTERDAM • BOSTON • HEIDELBERG • LONDON
NEW YORK • OXFORD • PARIS • SAN DIEGO
SAN FRANCISCO • SINGAPORE • SYDNEY • TOKYO

ELSEVIER

Academic Press is an Imprint of Elsevier

Academic Press is an imprint of Elsevier
30 Corporate Drive, Suite 400, Burlington, MA 01803, USA
525 B Street, Suite 1900, San Diego, California 92101-4495, USA
84 Theobald's Road, London WC1X 8RR, UK

Library of Congress Cataloging-in-Publication Data
Thorp, James H.
Field guide to freshwater invertebrates of North America / James H. Thorp,
D. Christopher Rogers.
   p.   cm.
ISBN 978-0-12-381426-5
1. Freshwater invertebrates–Ecology–North America. 2. Freshwater
invertebrates–North America–Identification.   I. Rogers, D. Christopher.   II. Title.
QL365.4.A1T44   2010
592'.176097–dc22

2010033213

British Library Cataloguing-in-Publication Data
A catalogue record for this book is available from the British Library.

ISBN: 978-0-12-381426-5

For information on all Academic Press publications
visit our website at books.elsevier.com

Working together to grow
libraries in developing countries

www.elsevier.com | www.bookaid.org | www.sabre.org

ELSEVIER   BOOK AID
             International   Sabre Foundation

Dedications from the Authors

"To my past and present graduate students who have inspired me and tolerated my frequent absence from the field while I wrote books."
James H. Thorp

"To my family and anyone else who has wondered what was going on under the water's surface."
D. Christopher Rogers

Contents

Part III
Ecology and Identification of Specific Taxa

Preface

Browsing through the shelves at bookstores found in colleges, commercial outlets, or state and national parks will reveal a host of field guides on birds, mammals, plants, and other nature subjects. But, it is the rare store that will carry a book on freshwater invertebrates, despite the joy many kids through adults derive from poking around streams, ponds, and wetlands and the importance of these aquatic ecosystems to nature and humanity. This is in part due to the attraction for nature lovers of large organisms like birds, the difficulty in finding many carefully concealed freshwater organisms, and the general lack of knowledge the public has on these organisms. Another important reason, however, is that invertebrate field guides for species other than terrestrial insects are extremely scarce—rarer than hen's teeth, as at least one of our mothers used to say. This dearth of books was the impetus for development of the *Field Guide to Freshwater Invertebrates of North America*. We believe that a greater focus by children through adults on freshwater invertebrates will expand the public's general knowledge and indirectly lead to improvements in the environmental quality of aquatic ecosystems and protection of endangered species. After all, people cannot protect what they do not understand.

Our field guide focuses on invertebrates that are large enough to identify at a general level using at most a hand-held magnifying glass. We include species of inland waters which are found in temporary pools to large lakes, water-filled roadside ditches to great rivers, ephemeral wetlands to riverside swamps, and surface creeks to cave streams. Strictly speaking, this book is not limited to "freshwater" invertebrates, because we also include species occupying inland saline habitats (e.g., brine shrimp in the Great Salt Lake) and crabs and shrimps migrating upstream from estuaries. Some of these organisms will be familiar to most people, such as crayfish, while others will be startlingly new to all but invertebrate specialists. We have avoided limiting coverage to "common species" because taxa abundant in one area can be rare or absent somewhere else, and this leads frequently to misidentifications. We cover species native to the USA and Canada as well as exotic species that have been transported purposefully or accidentally into our continent's inland waters. The level of identification in our field guide varies according to the group, size of the organism, and degree of knowledge required to classify the organism. For those readers wishing to gain further knowledge, we recommend that you consult advanced texts meant for college students through professional scientists, such as the third edition of *Ecology and Classification of North American Freshwater Invertebrates* (2010, edited by Thorp and Covich).

We hope that our field guide will be used by nature lovers everywhere, by students taught at home or in public schools, colleges, and universities, and by anglers looking for the best bait, lure, or fly.

Acknowledgments

Production of this book would not have been possible without the direct or indirect contributions of a number of people. In particular, we thank Alan P. Covich for all his work on the first three editions of *Ecology and Classification of North American Freshwater Invertebrates* (Thorp and Covich 1991, 2001, 2010, Academic Press). At the Elsevier publishing company, we are especially grateful to our acquisition editor Candice Janco, who was very supportive of the concept of this field guide, helped us with the design, and then "sold" it to the Elsevier book selection committee. From that point, the tireless effort of our development editor, Emily McCloskey, provided the vital assistance in taking the book from the word processing stage to the printed book. These are just two of the many fine people at Academic Press, Elsevier who made this book possible.

The process of producing this book would have been an order of magnitude more difficult without the contributions of hundreds of original photographs by Matthew A. Hill (see his picture on page xxvi) of EcoAnalysts, Inc. Matt volunteered his time to photograph invertebrates he had collected or obtained from colleagues throughout North America. Most of these were closeup shots of live animals, which made this process that much more demanding. We also appreciate our many other colleagues who contributed photographs of invertebrates in many phyla for this book. Their contributions are listed next to the figure numbers on page xxiii. We are also indebted to Gary Lester and EcoAnalysts for allowing Matt Hill to use the company's photographic equipment, collections, and laboratories.

The format of this book was based in part on Gregory C. Jensen's book Pacific Coast Crabs and Shrimp (1995, Sea Challengers Publications). Greg also contributed a photograph and was very encouraging of our project.

Many chapters in this field guide relied in part on a background of information from chapters in the *Ecology and Classification of North American Freshwater Invertebrates* (listed below as T&C III). In particular we would like to thank the authors listed below from whose chapters we extracted some information:

T&C III Chapter	Authors
2. An Overview of Inland Aquatic Habitats	James H. Thorp and Alan P. Covich
4. Porifera	Henry M Reiswig, Thomas M. Frost, and Anthony Ricciardi
5. Cnidaria	Lawrence B. Slobodkin and Patricia E. Bossert
6. Flatworms: Turbellarians and Nemertea	Jurek Kolasa and Seth Tyler
9. Nematoda and Nematomorpha	George O. Poinar, Jr.
10. Mollusca: Gastropoda	Kenneth M. Brown and Charles Lydeard
11. Mollusca: Bivalvia	Kevin S. Cummings and Daniel L. Graf
12. Annelida (Clitellata): Oligochaeta, Branchiobdellida, Hirudinida, and Acanthobdellida	Frederic R. Govedich, Bonnie A. Bain, William E. Moser, Stuart R. Gelder, Ronald W. Davies, and Ralph O. Brinkhurst
13. Bryozoans	Timothy S. Wood
15. Water Mites (Hydrachnidiae) and Other Arachnids	Ian M. Smith, David R. Cook, and Bruce P. Smith
16. Diversity and Classification of Insects and Collembola	R. Edward DeWalt, Vincent H. Resh, and William L. Hilsenhoff

Biographic Sketches of the Authors

JAMES H. THORP

Jim is a Professor and Senior Scientist at the University of Kansas' Department of Ecology and Evolutionary Biology and the Kansas Biological Survey. He has authored or edited six books in aquatic ecology and invertebrate zoology, along with over 90 journal articles and chapters. His research spans a great diversity of invertebrate types and focuses on community and ecosystem ecology. Jim has studied many freshwater and estuarine ecosystems, but he has emphasized river ecology for the last two decades. He is on the editorial board of two international journals and is currently President of the International Society for River Science.

D. CHRISTOPHER ROGERS

Christopher is an Invertebrate Zoologist at the University of Kansas and also the sole proprietor of Invertebrate Zoology Services, a consulting company specializing in identification of inland water invertebrates. He has published around 40 papers on invertebrate biology and is an expert on systematics and ecology of large branchiopods (fairy, tadpole, and clam shrimp). Christopher also conducts bioassessment research and consults extensively for US and foreign government agencies and companies. He has served on international editorial boards and is currently Associate Editor of the Journal of Crustacean Biology. Christopher is Vice President of the Southwest Association of Freshwater Invertebrate Taxonomists.

Photograph Credits

With the exception of the figures listed below, all figures were photographed and donated by Matthew A. Hill (see p. xxvi), using microscopes and cameras provided free by EcoAnalysts, Inc.

CHAPTER 2

Figs. 2a–i   Courtesy of D. Christopher Rogers.

CHAPTER 3

Fig. 3a, c-d, g, u, y, z	Courtesy of James H. Thorp.
Figs. 3e, f	Courtesy of Teresa M. Carroll.
Fig. 3b	Courtesy of Sarah J. Schmidt.
Fig. 3h	Courtesy of Walter K. Dodds.
Fig. 3i, p, r-t, ab	Courtesy of D. Christopher Rogers.
Figs. 3j, k	Courtesy of Horton H. Hobbs, III.
Figs. 3l–n	Courtesy of Craig E. Williamson.
Fig. 3o	Courtesy of Alan P. Covich.
Fig. 3q	Courtesy of Jo-szu (Ross) Tsai.
Fig. 3v	Courtesy of Kelly Kindscher.
Figs. 3w, x, aa	Courtesy of Robert L. Wallace.

CHAPTER 4

Figs. 4a, g, k	Courtesy of Chris Lukaup.
Fig. 4c	Courtesy of James H. Thorp.
Fig. 4h	Courtesy of Greg Jensen.

CHAPTER 5

Fig. 5a	Courtesy of Reiswig et al. (original photograph by the late Thomas M. Frost).
Figs. 5b, c	Courtesy of Reiswig et al.

CHAPTER 6

Fig. 6a	Courtesy of Peter Bryant.
Fig. 6b	Courtesy of Brigitte Galliot, University of Geneva.
Fig. 6d	Courtesy of K. Dunn.

CHAPTER 8

Fig. 8a   Courtesy of George O. Poinar, Jr.

CHAPTER 9

Fig. 9r   Courtesy of Robert T. Dillon.

CHAPTER 12

Fig. 12a   Courtesy of Timothy S. Wood.
Fig. 12c   Courtesy of Amee Baily.

CHAPTER 13

Figs. 13a, e   Courtesy of Jo-szu (Ross) Tsai.
Fig. 13h   Courtesy of Alison Smith.
Figs. 13i, l, p   Courtesy of John Pfieffer.
Fig. 13m   Courtesy of D. Christopher Rogers.
Fig. 13n   Courtesy of Christopher Lukhaup.

CHAPTER 14

Fig. 14c   Courtesy of D. Christopher Rogers.
Fig. 14d   Courtesy of Jeffrey S. Pippen at
                                 http://www.duke.edu/~jspippen/nature.htm.

CHAPTER 15

Fig. 15a   Courtesy of D. Christopher Rogers.
Fig. 15d   Photograph by the late Denton Belk,
                                 courtesy of D. Christopher Rogers.
Fig. 15h   Courtesy of S. Quinney.
Fig. 15j   Courtesy of Alan Longhurst.
Fig. 15m   Courtesy of Daniel G. Murrow.
Fig. 15o   Courtesy of Katherine Schnake.

CHAPTER 16

Fig. 16a   Courtesy of Robert Moeller.
Fig. 16d   Courtesy of John Pfieffer.
Fig. 16f   Courtesy of Mark Angelos.

CHAPTER 17

Figs. 17a, i, l, o   Courtesy of John Pfieffer.

CHAPTER 18

Figs. 18b, i, n	Courtesy of Chris Lukhaup.
Fig. 18e	Courtesy of Larry Serpa.
Fig. 18g	Courtesy of Ray Bauer.
Fig. 18h	Courtesy of D. Christopher Rogers.
Figs. 18j–m	Courtesy of Gunter Schuster.

CHAPTER 20

| Figs. 20a, c, d | Courtesy of Jason R. Neuswanger. |

Matthew A. Hill

The coauthors of this *Field Guide* acknowledge the contributions of hundreds of original photographs by Matthew A. Hill.

Part I

Introduction

Chapter 1

Using This Book Effectively

I. INTRODUCTION TO THIS FIELD GUIDE

A. What is the Nature of This Book?

Aquatic ecosystems contain an amazing diversity of invertebrates. Whether one is exploring a creek or river, a pond or lake, or even more unusual habitats such as vernal pools and water-filled tree holes and depressions in rock outcrops, you are bound to find bugs, wrigglers, rattails, snails, clams, mussels, elktoes, pocketbooks, mud bugs, crawdads, toe biters, hellgrammites, bloodworms, punkies, drakes, bloodworms, and many of varieties of small animals known collectively as invertebrates. This guide is designed to assist you in identifying the small aquatic animals that you may find living in freshwater (nonmarine) habitats.

Many of these organisms remain unseen due to their small size and secretive habits or are even ignored because identification is difficult. The invertebrates most commonly noticed are flashy or colorful (some crayfish, mussels, and beetles), good fishing bait (hellgrammites, caddisflies, mayflies), tasty (crayfish, shrimp, crabs), or an occasional painful nuisance now or later (toe biters, creeping water bugs, larval mosquitoes). However, with a little exploring, one can find an incredible array of aquatic invertebrates living under stream rocks, swimming in shallow pools, or hiding among plants in the margins of lakes and rivers.

Very few of our aquatic invertebrates can bite or pinch a human, and none can sting or poison you. Many of our aquatic invertebrates are important indicators of water quality and habitat health. All the water bugs, snails, crayfish, mussels, and worms have important functions in processing organic material, controlling algae, and feeding fish, birds, turtles, otters, and raccoons.

This book is divided into three parts. Part I (Chapters 1 and 2) introduces the organisms, provides information on collecting invertebrates, and describes the process of identifying invertebrates. Chapter 2 includes a key to help you determine which chapter you should read to learn more about a specimen you have collected. Chapter 2 also includes cautionary statements concerning protected species and protected areas where collecting is not allowed without specific permits. While reading these and other chapters, you may come across terms unfamiliar to you because of their scientific nature. To help alleviate some of these problems, we have provided a glossary at the end of this book. Part II (Chapters 3 and 4) is designed to inform the reader about the basic ecology and biology of aquatic invertebrates in general. This information will help in understanding the general importance of these creatures in the larger aquatic ecosystem, as well as where best to look for them, what their requirements for life are, and how they are used as ecological indicators. Finally, in Part III (Chapters 5–27) you can learn more about the diversity, distribution, form and function (including life history), ecology, and behavior of specific groups of invertebrates as well as how and where to collect them and techniques for culturing these organisms. These chapters also include photographs of specific taxa to help you identify the organisms you have collected. Keep in mind, however, that it is beyond the scope of this guide to identify all aquatic invertebrates to species level. The purpose of this guide is to allow the nonspecialist to identify aquatic invertebrates to an introductory level.

© 2011 Elsevier Inc. All rights reserved.
ISBN 978-0-12-381426-5, DOI: 10.1016/B978-0-12-381426-5.00001-6

B. Who Needs This Book?

Anyone who is curious about the animals that live in the creeks, rivers, ponds, lakes, and wetlands of North America will benefit from this book, although it is designed for the nonspecialist. Anglers may find this text particularly useful in identifying the critters that local fish are eating, both to use as bait or, for the fly fishing enthusiast, to identify models for wet flies. Many an angler has caught a fish and opened its stomach to see exactly which bugs the fish were hunting at that time of day.

Students in elementary, high school, and college classes may find this text useful for introductory natural history, entomology, zoology, or ecology lectures, field trips, and class projects involving aquatic ecosystems. Similarly, homeschool families can use this text as part of their curricula. Anyone wanting to learn more details about aquatic invertebrate ecology and biology as well as how to further identify aquatic invertebrates in general are encouraged to consult the most recent edition of *Ecology and Classification of North American Freshwater Invertebrates* (edited by Thorp and Covich in 2010) – the parent text of this guide. For help identifying aquatic insects in particular to much lower taxonomic levels (genus or species), you may wish to consult *An Introduction to the Aquatic Insects of North America*, edited by Merritt, Cummins, and Berg (2008).

C. What is an Invertebrate?

All living organisms are divided among five groups called kingdoms. These five kingdoms encompass the many species of bacteria (Monera), protozoa (Protista), fungus (Fungi), plants (Plantae), and animals (Animalia). In a very broad sense, animals are divided into two convenient groups: vertebrates and invertebrates. Only members of the former have backbones. Vertebrates are the fish, amphibians, reptiles, birds, and mammals. All other animals are invertebrates; in fact, about 90% of the identified species of organisms on this planet are invertebrates. Some examples of the latter are worms, leeches, snails, clams, insects, spiders, mites, crustaceans, sponges, and moss animals.

D. What Organisms are Covered in This Book?

This guide is designed to assist the user in identifying those freshwater aquatic invertebrates that can generally be seen in moderate detail with the naked eye, or at most with a hand lens. This book covers aquatic *macro*invertebrates and not *micro*invertebrates. The former are typically defined as invertebrates larger than 2 mm long, while microinvertebrates are of course smaller. Microinvertebrates include animals such as most nematode worms, many flatworms, wheel animals (rotifers), gastrotrichs, and protozoans. All these organisms can only properly be identified using a microscope.

Macroinvertebrates can be identified only into broad categories when using this book alone. Even though some of these animals are very large, the body parts that are needed for species level identification can often only be seen through a microscope and require dissection or special preparation. Moreover, proper identification typically requires a great deal of training and/or many more resources. Again, anyone wanting to identify aquatic invertebrates further than the scope of this guide should consult the most recent edition of Thorp and Covich's *Ecology and Classification of North American Freshwater Invertebrates*, published by Academic Press (Elsevier).

This field guide is primarily concerned with those creatures living on or below the water surface. However, some insects live in aquatic habitats as larvae or nymphs but migrate to terrestrial habitats as adults. These include dragonflies, mayflies, and some beetles and flies. Only the aquatic stages of those animals are extensively treated here. Similarly, the guide generally does not cover those macroinvertebrates living at the shore line but not in the water proper, such as tiger beetles and many spiders.

II. HOW INVERTEBRATES ARE CLASSIFIED

All organisms are placed into an internationally recognized system of classification in order to establish their relationships to each other. The classification system is hierarchical. This means that all the animals are divided into large groups, and then each large group is divided into smaller groups. Those smaller groups are further divided, until each specific kind is in its own category. This hierarchy is as follows, from largest to smallest:

> Kingdom
> > Phylum
> > > Class
> > > > Order
> > > > > Family
> > > > > > Genus
> > > > > > > Species

Each of these categories can be further subdivided if needed into "sub" and "super" categories, as in a suborder or a superfamily, but not all categories need this level of specification for all groups. Each category is called a taxon (plural: taxa), which means "name" in Latin. All the categories (taxa) except species are inclusive categories. This means that these taxa are composed of other smaller groups that share similar bodies or body parts. Only the species level category is exclusive.

All invertebrates are in the kingdom Animalia, which contains around 35 phyla (singular: phylum). Each phylum is defined by having a different body plan. For example, all animals with backbones are in the phylum Chordata. All other animal phyla contain invertebrates. Another example is the phylum Arthropoda, which is defined by having a body with a hard exoskeleton and jointed appendages.

To illustrate how organisms are placed into scientific groups, let us pick an organism, such as the common signal crayfish, to see how one generally classifies a species (see Chapter 2 for details on how to use taxonomic keys). This member of the phylum Arthropoda is native to the northwestern USA and adjacent Canada; but because of its popularity as a food item, it has been widely introduced across the northern and western parts of the continent.

The phylum Arthropoda is divided into several living subphyla: the Hexapoda (insects and springtails), Crustacea (crustaceans), Chelicerata (spiders, mites), and Myriapoda (millipedes, centipedes). Our crayfish belongs to the subphylum Crustacea, because it possesses two pair of antennae. The other subphyla possess either no antennae or at most one pair.

There are five classes in the subphylum Crustacea: Branchiopoda, Cephalocarida, Remipedia, Maxillopoda, and Malacostraca. The branchiopodans have leaf-like legs, so obviously our crayfish with its large claws and stout legs would not fit there. Cephalocarids and remipedes are only found in oceans and have more than 10 pairs of legs, unlike our crayfish. The maxillopodans do not have a telson, which is the central most piece in our crayfish's tail. Thus, our crayfish belongs to the class Malacostraca, whose members have large paired eyes and a strong abdomen with a telson.

The Malacostraca contains several orders. Six of these orders occur in our freshwaters. In the Cumacea, Isopoda, Tanaidacea, and Amphipoda, either the carapace (body shell) is absent or it does not cover the last part of the thorax. Members of the Mysida have seven leg pairs. Only in the order Decapoda, a carapace is present that covers the entire thorax and the five pairs of legs found in our crayfish. The word "Decapoda" literally means "10 legs."

Decapods are a large and diverse group that includes several infraorders (Caridea, Astacidea, and Brachyura) which occur in North American freshwaters. The Caridea contains the shrimp, separated from crayfish by laterally flattened abdomen. The brachyurans are the true crabs; they have an abdomen similar to that of a crayfish, but theirs is very flat and folded up underneath their body. All crayfish are in the Astacidea, with their straight abdomen that is flattened dorsoventrally (top to bottom).

We have now separated our signal crayfish from all other invertebrates, except other crayfish. Two crayfish families occur in North America: Cambaridae and Astacidae. Our crayfish is an astacid, separated from the cambarids by the males not having hooks on the lower parts of their legs. Cambarid crayfish have these hooks so that the male can hold onto the female during mating. The family Astacidae only has one genus in North America, *Pacifastacus*. This genus contains four species, but our crayfish can be separated from the others easily, because it has only one pair of spines on the tip of its rostrum (large plate that projects forward between the eyes). The other species have more spines, and some have patches of dense hair on their claws. Our signal crayfish lacks these characters, and so it is the species *Pacifastacus leniusculus*.

The species name is a binomial (literally, "two names"). It is composed of the genus name (*Pacifastacus*) and the specific epithet (*leniusculus*). The name *leniusculus* by itself is not the species name. Traditionally, the species name is always shown in italics, or if it is written by hand, then it is underlined.

III. A CAUTIONARY NOTE

A corollary purpose of this guide is to give the user an appreciation for the ecosystems where aquatic invertebrates are found. Many aquatic ecosystems are fragile, or have portions that are fragile in one way or another. For example, it sometimes takes months to years for a diverse community of diatoms, algae, and protozoans to grow on the upper surface of a rock, where the sunlight can reach them. Many invertebrates, including snails, mayflies, and caddisflies, rely on that aquatic garden for food. If one turns over the rock to look for invertebrates, the rock should be replaced so that the rich garden is still available for grazing invertebrates. If the rock is left upside down, the diatoms and other algae will die because no sunlight can reach them. Anyone investigating aquatic habitats should do so with as little disturbance to the habitat as possible.

Similarly, one should try to collect only what you need for your study and not overharvest the invertebrates. Many invertebrates can be cultured or kept in aquaria for observation and study. When the project is concluded, the living specimens should be returned to the water body from which they came.

As an important corollary to this, one should never release an invertebrate into a habitat other than the one where it was originally collected. This is especially true for snails, clams, mussels, crayfish, shrimp, crabs, and other organisms that do not fly. Many species invade new habitats due to human introductions, potentially causing terrible damage. The New Zealand mud snail has now become established in many parts of the USA and Canada, where it outcompetes native snail species that are important food for fishes. In contrast to native snails, this invasive species is difficult or impossible for our native fish species to digest. As other examples, the Louisiana red crayfish, the signal crayfish, and the northern crayfish were all introduced to California in the 1950s, driving one native crayfish species extinct and limiting another formerly widespread species to a couple of springs that are too cold for the invasive species. The zebra mussel, invading from the Ponto-Caspian region clogs water pipes, and just by the sheer numbers and weight of the accumulated individuals, kills crayfish and native mussels to which they attach. In 2002, the International Union for the Conservation of Nature identified nonnative invasive species as the number one threat to biodiversity worldwide.

We want to encourage people to investigate aquatic habitats and also, if possible, to attempt to recreate a small portion of the habitat under study in an aquarium. To enjoy aquatic organisms in an aquarium, you must ensure that the proper temperature, dissolved oxygen, water chemistry, and substrate are combined and functioning in harmony for the survival of its inmates. This forces the keeper to develop at least a rudimentary knowledge of ecosystem functions and hopefully, by extrapolating the effort required to maintain their aquarium microcosm to the larger rivers and lakes around them, develop a greater appreciation for the environment as a whole.

Chapter 2

General Techniques for Collecting and Identification

I. FINDING, COLLECTING, AND CULTURING AQUATIC INVERTEBRATES

A. Collecting Legally

The next two sections provide advice on what to do before you collect any invertebrates. They are meant to protect you—not scare you or dissuade you from exploring aquatic habitats.

Before collecting any organism, make certain that you have legal access to public and private sites because trespassing is punishable by law. The public is barred from some types of public lands, especially watersheds providing sources of municipal drinking water; these are generally protected for public health and safety reasons. It is also typically unlawful to collect in state and province parks, national parks and monuments, and wildlife refuges. However, permits are usually available for these areas. Often, all one needs is permission from the park manager or a ranger, and this is usually free. However, the fines can be large without this formal permission. Many U.S. states require permits before collecting invertebrates. In some states a fishing license is sufficient, whereas other states may require a permit only for selected organisms, such as crayfish or mussels. A fishing license is required to collect most crayfish species and bag limits may apply. Many states and provinces do not require children or K-12 classes to have collecting permits.

Permitting processes have been instituted to control the spread of potentially harmful and invasive species and to protect sensitive native organisms and habitats. For example, it is unlawful to collect the invasive Chinese mitten crab in California. Other species, like zebra and quagga mussels and some crayfish, are illegal to import live into some states; so if you are collecting in one state and cross into the next, please be aware of each state's regulations. Unless you have acquired specific permits, it is always illegal to collect threatened and endangered species. Many mussels, snails, fairy shrimp, crayfish, shrimp, and other species are legally protected. Even if you inadvertently disturb a protected species or their habitat, you could be prosecuted for violation of one or more endangered species laws!

Be informed about the legal conditions of any area where you want to explore. Different counties, states, provinces, and nations have different laws, often for different reasons; do not assume the laws of one area apply to another. The more you know, the better protected you will be from inadvertent, unlawful activities and potential fines, jail sentences, or (as once the junior author suffered for trespassing) shotguns filled with rock salt.

B. Collecting Safely

Any activity near or in water can be potentially hazardous. One should always exercise normal, rational caution around aquatic habitats, especially creeks and rivers. Swift water can quickly carry a

ISBN 978-0-12-381426-5, DOI: 10.1016/B978-0-12-381426-5.00002-8

person away if they accidentally slip into the current. Creeks and rivers may be filled with slippery rocks. In rocky areas, maintain a stance with your feet well separated for more stability; felt bottomed boots or waders help sustain a grip on slippery substrates. Always wear a life vest when using a boat, canoe, or kayak or when wearing chest waders. Such personal flotation devices are also good for working in or near swift, cold water. Rapids can be very dangerous; and if you fall into the current, you may be beaten against rocks or logs. Swift currents (including around low-head dams) and waterfalls can draw even the strongest swimmer under water and pin them to the bottom. Flowing waters can undercut banks, making them liable to collapse when you stand upon them. Flash floods are especially common in arid regions such as the Southwest, where distant rains can swell a deeply channelized creek from a few centimeters to a few meters deep in a matter of minutes. Because of water current flow paths, areas less than a meter from shore can be very shallow at one site and over your head a few meters away. Muddy shore lines and lake bottoms can be more easily traversed in "coot boots," which will help prevent you from becoming mired. Similarly, quicksand can trap the unwary. If you find yourself immersed in quicksand, the best strategy for escape is to lie out as flat as possible and swim to the safety of the shore or firm substrate.

Some animals found in or near water bodies can be dangerous in some regions of the USA. In the East and Southeast, the water moccasin (or cottonmouth) snake can be a hazard in streams, lakes, and wetlands. These snakes are very poisonous and are our only dangerous water snake. However, many other nonpoisonous water snakes are mistaken for this reptile, some of which mimic the moccasin when aroused. Cottonmouths often lie on branches and logs at or well above the water surface and may be found in shoreline grasses. They are excellent swimmers and normally avoid humans, but they can be fairly aggressive if irritated. Snapping turtles occur in eastern North America from southeastern Canada, south to Florida and through eastern Texas. These animals are usually shy and retiring; but if they feel threatened or cornered, they can give a wicked bite. Typically they hiss and rock back and forth in an attempt to scare off any threat. The largest species is the alligator snapping turtle, which is an endangered species. The American alligator occurs in the southeastern USA. This large "lie in wait" predator does an excellent job of waiting motionless until a prey item comes close enough for attack. A hungry alligator may also pursue prey. Females that are brooding eggs or young on shore are especially dangerous. Florida has become home to many nonnative invasive species including the Burmese python. These large snakes are established in the southern, tropical portion of the state and have killed alligators and deer as well as one small child. In the Pacific Northwest, rivers and streams are frequented by grizzly bears, particularly during salmon runs. Water courses with actively fishing grizzlies should be avoided . . . unless you want to expand their diet!

On the opposite end of the animal size scale are some small aquatic creatures that are mostly just pests but which can occasionally transmit diseases. Blackflies, biting flies, and mosquitoes are in both categories. However, mosquitoes are of special concern throughout most of North America, spreading encephalitis (including west Nile virus) and malaria. These diseases are especially prevalent in California and the Gulf Coast states. Although these diseases are extremely rare now for people who confine their travels to the USA and Canada, travelers to tropical regions need to be very cautious. Leeches and flukes (trematodes) are irritants in many regions, but these rarely cause serious problems in North America.

These lists of potential hazards are by no means exhaustive. The cardinal rule, therefore, is to be careful and cautious and not take risks.

C. Finding Aquatic Invertebrates

Aquatic invertebrates are relatively easy to find. In flowing water and wave-swept, rocky lake shores, one can typically find a great variety of invertebrates on, among, and under rocks. Vegetated shore lines and pool margins are refuge for many invertebrates as well. Submerged or floating logs are often good places to hunt. Cracks and crevices in wood can hide many small critters. Many invertebrates live in sand bars. Some mayfly nymphs, caddisfly larvae, and clams live in the shifting sands. Mayflies can be very difficult to catch as they move quickly away from any disturbance. Leaf packs and woody debris dams in flowing water also harbor a variety of invertebrates. Overhanging banks, tree limbs, and vegetation also provide cover for invertebrates. Exposed hard surfaces, such as dock pilings, boat bottoms, dam faces, and rocks, provide attachment sites for sponges and bryozoans as well as homes for caddisflies, midges, snails, and some mussels.

Current velocities, temperatures, and oxygen conditions influence the distribution of many species. In flowing waters, the most oxygenated portions are the areas just below riffles. Because oxygen levels are high, more invertebrates congregate here than in other parts of a creek or river. Many predatory fish typically wait below riffles for any bugs that get carried away by the current.

The prevalence of easily collectable invertebrates varies with time of day. Many invertebrates, such as crayfish, shrimp, and some snails, are nocturnal. Many crayfish live in deep burrows or in the deepest portions of rivers during the day and only emerge into shallow areas at night. In many cases this seems related to predator avoidance.

D. Collecting Aquatic Invertebrates

There are numerous ways to collect aquatic invertebrates. Professional zoologists and ecologists have many specialized traps, dredges, samplers, nets, and grabs designed for gathering specimens. However, for the average person there are easier and less expensive ways to find and collect aquatic invertebrates. Our individual taxonomic chapters provide more detailed information, but you will find below some general advice.

Far and away the easiest collecting technique is to use a shallow bucket or plastic pan. First, fill a bucket or pan with water from the habitat being explored. Then, use your hand or a soft brush to remove seen or hidden invertebrates from the rocks and sticks and place them in your bucket or pan. You can also place aquatic plants or handfuls of gravel and sand into the bucket or pan; the small invertebrates will soon become visible as they move about the container. White plastic is usually best, as the invertebrates show up against the bright surface.

A simple aquarium dip net is convenient for collecting specimens, but in a pinch, a plastic kitchen sieve with a handle will work. The average aquarium dip net comes in a variety of sizes, with the largest (10 inch diameter, or about 25 cm) being the best. The handle is typically made of a soft metal folded into a terminal loop that is plastic coated. The handle loop can be easily flattened slightly, allowing it to fit snuggly inside the end of a 1 inch diameter PVC pipe (Fig. 2a). This pipe can be purchased in varying lengths to add reach to your collecting endeavors. The net can be easily used to sweep marginal vegetation, or to root around under overhanging banks and vegetation, or to sift sand (Figs. 2b–c). The smaller the net mesh size, the more you will catch.

FIG. 2A. Dip net and PVC pipe handle.

FIG. 2B. Collecting from shore.

FIG. 2C. Sweeping through vegetation in a wetland.

Some aquatic invertebrates drift with water currents to escape local disturbances from a predator or competitor or to seek habitats with more abundant food. This process is called "drift" and it may be accidental or intentional, occurring more often at night. Unintentional drift occurs most frequently in areas of higher turbulence and when the stream is flowing rapidly. This behavior can be exploited by placing a net just downstream of a riffle or log jam to passively collect the drifting organisms (Fig. 2d). Alternatively, you can be the disturbance force by kicking around loose rocks (usually cobble or larger) and sticks (Figs. 2d–g) to cause the invertebrates to drift into the waiting net!

Somewhat larger invertebrates can be collected by other means. Crayfish and some shrimp can be captured readily with any good commercial crayfish trap using appropriate bait. Mussels can be turned up by pulling a rake gently across the gravel bottom of a shallow headwater stream. Please remember, most mussel species are very rare and should not be disturbed. Also, most mussels and crayfish are protected by law and either a fishing license or special permit may be required to collect them depending upon your location.

Finally, another useful way to collect aquatic invertebrates is a simple aquatic light trap (terrestrial light traps are used to collect flying insects) (Fig. 2h). All you need is a small bucket with a lid, a

FIG. 2D. Collecting in a riffle area of a stream.

FIG. 2E. Examining a rock for invertebrates.

rope (if the water is deep), a rock, and a battery-operated glow stick (chemical sticks are less preferred because they can be used only once). This trap is deployed at dusk or just after dark. The rock is placed in the bucket. It must be heavy enough to keep the bucket from floating away. The glow stick is activated and hooked under the rock so that it will not float away and will stay on the bottom of the bucket. The bucket lid, with two or more large slot shaped holes in it, is placed on the bucket. Fill the bucket with water and place it on the bottom of the stream, pond, or pool to be sampled. If the water is deep, attach a rope from the bucket handle to an easily retrievable place on shore. Many insects, mites, and crustaceans are positively phototropic. This means that they will be attracted to the light of your trap and enter the bucket through the holes in the lid. After a few hours pull up the trap and examine the contents. A trap left out overnight may lose specimens, which will swim out when they see the morning light. Large, predatory specimens can be captured and may reduce the remaining collection to carnage!

E. Records Keeping

One should always keep good records of where and when specimens were collected (Fig. 2i). This information helps us to understand the organism's habits, habitat, and activity periods. A diary recording a day's collecting should start with a description of where the collecting took place, the time and weather, a habitat description, and a list of organisms observed and collected. The locality data should be in the following format:

Country
> State, Province, or Territory
>> County, Parish, or Regional District
>>> Locality description

In this way, your data go from the larger to smaller area. This helps someone else using your data to find a location easily and in a logical format. For example:

USA: CA: Shasta County: Poison Lake at intersection of railroad tracks and Pittville Road, north of Highway 44.

GPS or latitude/longitude coordinates can be added to this information to make the local site easier to find. These same data should be placed on collection labels for individual specimens, followed by the date of collection, in the following format: day/month/year. For example:

2JUN2009 or 2VI2009

This abbreviated format uses the least characters and thereby the least space. This is important when there is so little space on the specimen label for writing. Following the date, the collector's name (initial followed by last name) ends the label data, unless one uses catalog numbers to correlate the specimen with other records or a collection catalog.

F. Preserving Specimens

Simple techniques are available if you need to preserve a specimen for further study. The best preservative is ethyl alcohol (sometimes abbreviated as ETOH). This can be obtained inexpensively from most pharmacies as one form of rubbing alcohol. Another easily obtained preservative is isopropyl alcohol. Both will work, but the ethyl alcohol is required for any later genetic studies.

First, preserve your specimens in the alcohol, which will draw the excess water from the specimen's tissues. Second, replace the initial alcohol with fresh alcohol after 24 hr. If there is a large amount of organic material in your preserved sample (leaves and sticks), or more specimens than alcohol, the preservative should be changed after 6 hr and again after 24 hr to prevent later decomposition (rot) of the material.

Museum quality specimens should be placed in a glass vial with alcohol,

FIG. 2F. Looking through woody debris.

FIG. 2G. Examining the net for captured invertebrates.

FIG. 2H. Aquatic light trap for collecting larval and adult invertebrates.

filled nearly to the top. Place inside the vial or bottle a label written with a pencil or an alcohol-proof pen that includes the information described in Section I.E. Laser jet printers can be used to make labels; however, once the labels are printed, they should be heated in a microwave oven for a few minutes. Laser jet printers place the characters on the paper using wax, which will dissolve in alcohol, thus causing all the characters to fall off the paper and lie in a jumble in the bottom of the vial after a few months.

G. Culturing Invertebrates in an Aquarium

Most aquatic invertebrates can be kept in an aquarium, provided that their needs for dissolved oxygen, current, food, refuge, and waste removal are met. The best aquariums are the largest, with an abundance of live plants. The plants help oxygenate the water, absorb animal metabolic wastes, and keep the water clear. Sometimes filtration is needed as well. Similarly, it is best to exchange weekly a third of the aquarium volume with rainwater, or better yet, water from the habitat where the specimens originated. The larger and better functioning the aquarium, the less frequently the water will need changing. Tap water should not be used unless it has been kept standing in open air for a day or so, so that the toxic chlorine can escape.

Maintaining appropriate concentrations of dissolved oxygen is essential. Most aquatic invertebrates obtain oxygen dissolved in the surrounding water through gills or across the general body surface, but a few invertebrates collect air bubbles from the air–water boundary. The basic rule is that the cooler the water, the more dissolved oxygen it contains. Unfortunately, it is difficult to keep aquarium temperatures cool enough to meet the dissolved oxygen requirements of most flowing water invertebrates. The addition of one or more air pumps can be helpful.

FIG. 21. Examining part of the catch.

Water current needs can usually be satisfied with the use of a power head or a similar water pump. Filtration, if needed, should be accomplished by an outside filter or a sponge type filter, never an "under gravel" type filter. These filter types will promote very healthy root growth, but the plants will suffer.

Providing the appropriate type and amount of food (but not too much) is often a major challenge. Predatory species need other invertebrates to feed upon, and their remains may foul the tank. Grazing species need algae and periphyton to scrape from rocks and other hard surfaces, or plants to browse upon.

Refugia are equally important. Many snails and flatworms are nocturnal and need dark daytime hiding places. In a community type aquarium, potential prey animals need to be able to hide from predators or else the aquarium becomes a slaughterhouse. Some organisms, such as crayfish, require individual shelters to avoid damage from frequent fighting.

Easy species to rear in aquaria include snails, diving beetles, water bugs, leeches, crayfish, and shrimp. All but most snails need a tight-fitting glass or screen lid on the tank, or else they may escape. Many insects can fly directly from the water surface.

II. HOW TO USE A DICHOTOMOUS KEY

All invertebrates can be identified to lower taxonomic levels with a dichotomous key (or in some of the older texts, a table of keys), which is a guide or tool used to identify an organism by a series of paired sets of opposing character states. These character states are mutually exclusive. The key is arranged in a series of couplets, numbered sequentially. Each couplet is composed of two rubrics.

To identify a specimen, one starts with the very first couplet in a key. The user must decide which of the two opposing character states is found on their specimen. By way of example, the first couplet in the key found in Section III reads:

1   Legs absent 2
1'   Legs present ....INSECTS, CRUSTACEANS, MITES, AND SPIDERS ....Chapter 13

In this example, if your specimen is lacking legs, you continue to couplet 2. If it has legs, you go directly to Chapter 13, and use the key there to identify your specimen further. Couplet 2 looks like this:

2(1)   Animals in colonies, or clumps, or large masses, attached to rocks, sticks or other hard
         surfaces 3
2'   Animals separate, not in colonies 4

Again, you have a choice as to which character state best describes the specimen that you are attempting to identify. The "(1)" tells you what couplet you used to get to the present couplet, in case you need to backtrack. In this instance, depending on the growth form of your specimen (colonies or individuals), you would then proceed to either couplet 3 or 4.

Ultimately, you will reach a terminus in the key. The termini will provide a name for the organisms at the level of discrimination in the key (which varies among organisms and keys). In the case of the termini in Section III, you are directed to the proper chapter in the book for the next stage of identification.

III. INTRODUCTORY TAXONOMIC KEY FOR INVERTEBRATES AND CHAPTER GUIDE

1   Legs absent...2
1'   Legs present.................INSECTS, CRUSTACEANS, MITES, AND SPIDERS.......Chapter 13

2(1)   Animals in colonies, or clumps, or large masses and attached to rocks, sticks,
        or other hard surfaces...3
2'   Animals separate, not attached to hard surfaces...4

3(2)   Animals forming dense mats or finger like projections, without obvious moving body parts,
        similar to algae, but crusty in texture, usually green or brown........SPONGES..........................
        .... Chapter 5
3'   Animals forming colonies that are branched or jelly like, with moving fans that sweep the
        water...MOSS ANIMALS..................Chapter 12

4(2)   Body in a shell..5
4'   Body without a shell..6

5(4)   Shell in one part, as a spiral or a low cone, head always present ....
        ....SNAILS AND LIMPETS............Chapter 9
5'   Shell in two equal parts, eyes and head always absent...
        ....CLAMS AND MUSSELS.........Chapter 10

6(4)   Body segmented..7
6'   Body not segmented..8

7(6)   Body with more than 15 segments, may have a sucker at one or both ends, never with an
        obvious head................................SEGMENTED WORMS AND LEECHES..........Chapter 11
7'   Body with less than 15 segments, never with suckers, almost always with an obvious head
        ....INSECTS.................Chapter 19

8(7)   Body without tentacles..9
8'   Body with a ring of tentacles................................HYDRA AND JELLYFISH..........Chapter 6

9(8)   Body round in cross section, extremely long and thin ....
        ....HORSEHAIR WORMS...........Chapter 8
9'   Body flat in cross section..FLATWORMS ....Chapter 7

III. INTRODUCTORY TAXONOMIC KEY FOR INVERTEBRATES AND CHAPTER GUIDE

Part II

General Ecology of Freshwater Invertebrates

Chapter 3

The Nature of Inland Water Habitats

I. INTRODUCTION

Inland water habitats of North America include lentic and lotic ecosystems on the surface (=epigean) and underground holding fresh, acidic, saline, and alkaline waters. Although the percentage of the planet's total biospheric water contained within inland habitats is very small (1%), it is vital to the inland aquatic biota, terrestrial life, and human populations. Lentic habitats include ephemeral pools, permanent wetlands, roadside ditches, natural ponds, and fresh and saline lakes as well as artificial systems from small cattle troughs and farmers' ponds to large hydroelectric reservoirs. Lotic ecosystems range in size in North America from intermittent streams to the mighty Mississippi River. Although most inland waters are held within large ancient lakes, such as the Great Lakes of Canada and the USA, the greatest diversity and densities of invertebrates and fish occur in smaller lakes, wetlands, creeks, and rivers. These modest ecosystems are more important to diversity because their overall habitat complexity is much greater and a much larger percentage of their volume is in the shallow waters of the photic zone. Most invertebrates reside and obtain their food in the photic zone, which is defined as having light intensities of 1% or more of surface radiation.

The distribution and abundance of invertebrates are strongly influenced by physical and chemical habitat characteristics and both positive and negative species interactions. This chapter introduces key physicochemical features of inland water habitats, while Chapter 4 briefly discusses ecological relationships. For more information on these subjects, see "Chapter 2: An Overview of Inland Aquatic Habitats" in Thorp and Covich (2010) and many stream and lake ecology books, such as Wetzel (2001), Dodds (2002), and Allan and Castillo (2007). Chapters 3–4 of our field guide draw extensively from that chapter.

II. LOTIC ENVIRONMENTS—CREEKS AND RIVERS

Many freshwater invertebrate taxa are restricted to headwater streams and rivers by their unique environmental characteristics, and fish and invertebrate diversities are substantially greater in lotic than lentic habitats. This reflects in great part the greater average evolutionary age and habitat diversity of streams (Figs. 3a–d).

Streams are typically more turbulent than lakes and, therefore, their waters rarely stratify into the distinct vertical thermal, nutrient, and chemical layers found in lakes. Oxygen levels are typically higher in streams. Winter ice is usually thinner than in lakes and rarely extends below a surface layer. [Note: while it may be safe to walk or skate on lake ice in some areas and on some occasions, avoid doing so on rivers because the ice thickness is both deceptive and highly variable in time and space.] Flowing water habitats are heterogeneous, especially when you include vertical (hyporheic zone within the sediment) and lateral components of the riverine landscape (main channel, slackwaters, and floodplains).

ISBN 978-0-12-381426-5, DOI: 10.1016/B978-0-12-381426-5.00003-X

FIG. 3A. Constricted channel Colorado River.

A. Longitudinal and Lateral Changes in River Structure

Many changes in the habitat template occur longitudinally (upstream to downstream) and laterally (main channel to side channels and floodplain pools) from headwater creeks to great rivers. Most rivers merge downstream with other rivers or the sea and in the process typically widen, deepen, increase in average velocity, and decrease in turbulence. Species distribution patterns occur at all scales from the large basin/watershed level (e.g., fish diversity increases with stream size and permanence in general), to the valley level where channel form may change dramatically, and to smaller spatial scales of a reach (e.g., a repeating unit of riffles and pools) or even macrohabitat like a rock or wood snag. Flow within macro- and microhabitats can be substantially lower or higher than the average current velocity. Side channels and backwaters have much lower and in some places zero flow. In general, smaller spatial scales are more pertinent to distributions of individual species.

Streams are sometimes roughly classified by their order, or pattern of tributary connections. Thus, you may read that a species frequents 1st to 3rd order headwaters, 4th to 6th order creeks and small rivers, or 7th order and higher, medium to large rivers (by comparison, the Mississippi River reaches

FIG. 3B. Riffle and pool areas in the Pecos River in a semi-arid region of New Mexico.

FIG. 3C. Riffle area in the Mianus River of Connecticut, which despite its name, is a headwater stream in this area.

at least 12th order). This is a simple technique which aids analysis of longitudinal changes in stream characteristics within a single catchment; however, it is often misleading when comparing streams in different ecoregions, such as those in eastern humid forests versus arid southwestern ecoregions, where a raging torrent can replace a dry stream bed in a few minutes.

FIG. 3D. Upper Three Runs Creek in South Carolina. Note the large amount of submerged macrophytes in this sand-bed stream.

B. The Importance of Substrate Type and Natural Flows

Along the river's usually sinuous path, one can find alternating sand or gravel bars, pools, and riffles, sometimes accompanied by multiple channels and oxbows (cut-off channels). A sequence of riffles, runs, and pools can occur within a few hundred meters. Sediment is continually being eroded from some areas (e.g., undercut banks and the head of islands) and deposited in other regions (such as on bars and the foot of islands).

Substrate type and size are of prime importance to survival of the mostly benthic invertebrate populations. Substrates provide sites for resting, food acquisition, reproduction, and development as well as refuges from predators and inhospitable physical conditions. Hard substrates are generally valuable only for their surface qualities (except for wood-boring species), whereas silt to gravel substrates provide a three-dimensional environment. A substrate's desirability is species specific and based on whether it is mineral, living, formerly living, small, large, and stable or more mobile. Its importance is also influenced by the physical–chemical milieu in which it resides (e.g., depth and the surrounding current velocity and oxygen content). In general, well-flushed, stony riffles support more invertebrate species and individuals than silty reaches and pools, as a partial consequence of the different adaptations required to live in these physically distinct areas. Shallow lotic and lentic waters tend to have more species and greater average densities. Wood snags are ecologically vital in many streams, especially in sand-bed rivers where other natural hard substrates are rare. Unfortunately, governments have been removing fallen trees from navigable rivers since the nineteenth century.

Streams are characterized by fluctuating depths, velocities, and discharges in contrast to the relative depth constancy in lentic systems other than ephemeral pools. The smaller the stream and the drier the ecoregion, the more variable and unpredictable are a stream's discharge. The importance of the natural flow regime has become almost a paradigm in stream ecology, but stream flow is highly regulated in most rivers to the detriment of native organisms.

When stream current velocities rise, many invertebrates seek substrate shelters. This is generally easier for invertebrates in well-mixed rocky substrates than in finer substrates, which can be difficult to penetrate and subject to low oxygen levels. Keep in mind, however, that most species normally reside in habitats with lower than average stream velocities, though filter-feeding animals may select high-velocity areas. Many stream invertebrates are partially protected from dislodgment by having a low vertical profile. They also may be characterized by streamlining, reduction of projecting structures, development of holdfasts (suckers, friction pads, hooks, or grapples), and adhesive secretions.

C. The Thermal and Chemical Environment of Streams

Species' distributions, abundances, and emergence patterns are correlated with the annual mean and range of temperatures as well as with smaller-scale thermal effects from groundwater inflows and canopy shading. While temperature constancy increases with stream size, the downstream pattern in average temperature is greatly affected by flow direction, overall stream bed slope, presence of springs and canopy coverage in headwaters, and location of reservoir dams and their release pattern.

Natural chemical features of inland waters affect invertebrate abundance and diversity, especially oxygen levels, water hardness (salinity), and sometimes hydrogen ion concentration, or pH. Pollution from humans severely impacts ecosystem integrity by affecting these chemical parameters as well as by introducing organic and inorganic toxicants.

Dissolved oxygen levels tend to decrease downstream because of lower turbulence, often higher temperatures, and greater microbial demands from organic decomposition. Low oxygen levels are more frequently encountered in lakes, seasonally sluggish streams, and river backwaters than in fast-flowing lotic habitats.

Inland waters are usually extremely dilute compared with seawater, with four cations (calcium, magnesium, sodium, and potassium) and four anions (bicarbonate, carbonate, sulfate, and chloride) comprising almost all ions in inland water systems. "Soft waters" are generally found in catchments with acidic igneous rocks, whereas hard waters occur in watersheds with alkaline earths, usually derived from limestone and other calcareous rocks. North American waters tend to be slightly harder on average (142 mg/L) than the global mean salinity of river water (120 mg/L). Animals with either partially calcareous exoskeletons (e.g., crayfish, seed shrimp, and water fleas) or hard calcareous shells (snails and mussels) have added needs for calcium ions and are sometimes excluded from soft water habitats.

Most natural waters tend to be somewhat acidic because rainwater is naturally slightly acidic. Streams in forested basins tend to be more acidic, especially those with large stands of coniferous trees which contribute humic acids to the waters. Bogs and the streams draining them are among the most naturally acidic ecosystems, with some sphagnum bogs having pH values below 4.5. Streams draining limestone basins are well buffered by carbonates and, therefore, have neutral to high pHs. The pH of streams draining coal and other mining areas are often lethally acidic.

D. Intermittent Streams and Spring Systems

Intermittent streams fluctuate regularly from dry to wet (Figs. 3e–h), forcing resident inverte-brates to move upstream (if spring-fed) or downstream, seek shelter in deep pools, burrow within sediments, emerge as adults (some aquatic insects), develop resistant stages, or lay eggs and then die.

While perhaps most headwater streams are fed by terrestrial runoff, many originate from cold or, less frequently, thermal springs. Near the spring source, temperatures in these streams fluctuate less and oxygen concentrations are occasionally lower, sometimes to near lethal levels for invertebrates. A few invertebrate species survive in areas of hot springs where ambient temperatures may approach 40°C (Fig. 3i). Waters in geothermal springs also may have high sulfur concentrations. On the other hand, springs can be important refuges in headwaters of the Great Plains and other arid regions where surface waters downstream periodically disappear. Biota of springs often varies from other first-order streams as a result of their connection to fauna of underground rivers. Because springs are relatively uniform environments over long ecological periods, it is not unusual to find relict species in these environments that have survived the retreat of the glaciers only in these limited habitats.

FIG. 3E. Intermittent stream in the Ozarks of southeastern Missouri in high water condition.

FIG. 3F. The same stream as in Fig. 3e but all the surface water has disappeared, leaving only subsurface (hyporheic) aquatic habitats.

FIG. 3G. Intermittent tributary of the Darling River in Australia which dries to isolated pools.

FIG. 3H. Intermittent prairie stream in Kansas with a spring fed, headwater pool which provides a haven for some aquatic organisms when downstream areas dry.

III. SUBTERRANEAN HABITATS: HYPORHEIC ZONE AND CAVE ENVIRONMENTS

Although almost all invertebrates one will collect are found from the air–water interface down to the silty, sandy, or rocky "bottom" of the stream, many very small species live within the stream sediment in an area called the hyporheic zone (a somewhat comparable term in lakes is the psammon zone). The hyporheic thickness varies greatly with bed topography, substrate porosity, and water velocity,

FIG. 3I. Hot spring area in northern Nevada which gives rise to a thermal stream.

FIG. 3J. Subterranean aquatic habitat in southern Indiana with a small stream slowly cascading down a limestone staircase.

and is thus greater in rivers with coarse than fine substrates. The overlying surface water (epigean) affects and is influenced by the temperature, oxygen level, and nutrient content of hyporheic waters. The hyporheic community can consist of large individuals or small interstitial species depending on the physical conditions. Almost all freshwater invertebrate phyla have representatives in hyporheos, except possibly Porifera (sponges) and Cnidaria (hydra).

Subterranean cavities often develop in regions with large amounts of soluble limestone and adequate underground water. This karst topography, which is widespread in North America, features both surface and underground streams which emerge through cave openings as karst springs with distinct invertebrate communities (Figs. 3j, k). Food in the subterranean ecosystems is derived from surface streams or wastes from animals entering the cave, such as bats. Food webs are relatively simple, consisting mostly of detritivores and their predators. Temperatures are very constant and light is absent or dim at best. Abiotic disturbances are less frequent and severe, though flooding occurs following strong surface rains.

Species diversities are low in subterranean ecosystems with a high rate of endemism from the forced isolation over thousands of years. Indeed, 50% of the 1000 aquatic invertebrate species in North American caves are found in only 18 counties, and 95% of the cave species are listed by the Nature Conservancy as

vulnerable or imperiled. In comparison to surface species and facultative troglophiles, the obligate troglobites (or troglobionts) usually have lower metabolic rates, much longer lifespans, and lengthened reproduction. They may also be blind and lack body pigmentation.

IV. LENTIC ECOSYSTEMS

A. Lakes

FIG. 3K. A blind cave crayfish (*Orconectes inermis testii*); note the lack of body pigmentation (the red spot is a dye marker to keep track of individuals) and the absence of eyes.

Lentic ecosystems consist of all non-flowing, surface water bodies not directly connected to a river. They range from ephemeral pools to ancient lakes and from shallow, heavily vegetated wetlands to deep lakes with minimal aquatic macrophytes. Most contain fresh water, but saline pools and lakes are common in the arid West. Natural lakes (Figs. 3l–o) form from both catastrophic phenomena (landslide, glacial, tectonic, and volcanic activities) and less violent actions of rivers (channel cutoffs), waves, and rock solution. Even animals such as beavers participate in pond formation. North America has more lentic ecosystems per hectare than any other continent because of glacial activity in the last 20,000 years. Relatively permanent lakes are most common in Canada and northern USA where glaciers formed, but they are relatively rare in other regions of the USA. Human activities produce lentic systems from deep reservoirs to shallow constructed water troughs for livestock to ephemeral pools formed in wet soil by careless automobile drivers. In areas like the Great Plains, the importance of ephemeral wetlands has been reduced by the construction of thousands of farm ponds and large reservoirs. Most natural and human-constructed lentic systems have average depths <20 m.

FIG. 3L. Lake Lacawac, a pristine glacial lake in Pennsylvania.

FIG. 3M. Emerald Bay on the western (Californian) shore of Lake Tahoe in formed by faulting in the earth crust; the lake is the second deepest in the USA.

FIG. 3N. Lake O'Hara, a pristine glacial Lake in the Canadian Rockies which is noted for nearby Burgess Shale fossil deposits.

Lentic ecosystems are divisible into several abiotic zones, primarily based on distance from shore, light penetration, and temperature. Lakes can be divided into the aphotic zone (<1% light penetration) and the photic zone, with the latter including the open water pelagic (or limnetic) zone and the nearshore littoral zone where rooted macrophytes can grow. Temperatures fluctuate more rapidly in shallow water,

FIG. 3O. Chain of tundra lakes near the Toolik Lake Field Station close to the foothills of the Brooks Range in Alaska's North Slope.

and wave turbulence is greater. Oxygen is rarely limiting nearshore (except sometimes in eutrophic lakes), but the effects of ice formation can be more severe.

Most permanent lentic ecosystems in North America become seasonally stratified with a lighter layer of water, the epilimnion, floating over a denser hypolimnion. The metalimnion separates the two layers and is characterized by an area of rapid temperature change called a thermocline. The epilimnion is commonly warmer in summer and colder in winter than the hypolimnion. Because water exchange is minimal between upper and lower zones during stratification, the hypolimnion is frequently much lower in oxygen, higher in nutrients, and different in pH and other chemical concentrations. When lake stratification breaks down for a short period during one or more seasons, much or all of the water mass recirculates during lake turnover.

Lake invertebrate assemblages can be subdivided primarily into "zooplankton" (mostly protozoa, rotifers, cladocera, and copepods) within the water column and "benthos" which live in, on, or just above the bottom (commonly in association with macrophytes). A third, smaller group encompasses the "neuston" which live at the air–water interface. Holoplankton live their entire aquatic lives within the water column, whereas meroplankton (e.g., early instar midges) spend only a limited time there. Some species live in deep water during the day and then migrate at night to shallow waters. A less-pronounced pattern occurs from littoral to pelagic areas over the same period. Most freshwater benthos reaches the maximum densities and diversity in shallow water and decline perceptibly with increasing depth in the profundal zone. Many benthic species are infaunal (living in the substrate), others crawl upon the bottom inorganic or organic surface, and many perch or climb among vegetation in the littoral zone.

B. Wetlands, Swamps, and Ephemeral Pools

For many years the public has ignored or considered undesirable the vast acreage of wetlands, swamps, and ephemeral pools, considering them breeding grounds for mosquitoes and snakes. Many have been drained, filled, and bulldozed for housing and commercial development without regard to their intrinsic value to wildlife and the environment. We now know that they are critical to resident and migratory animals and are often vital in filtering pollutants from the environment. While our knowledge of these systems is relatively limited compared with other inland water habitats, strict environmental laws have

been adopted in the last quarter century to protect these habitats. The term "freshwater wetlands" refers
to nontidal ecosystems whose soils are saturated with water on a permanent or seasonal basis. Emergent
aquatic vegetation is prominent and may alternate with annual terrestrial plants in these marshy habitats.
Wetlands vary from shallow, seasonally ephemeral pools (e.g., rocky outcrop pools, Carolina bays,
prairie playas, vernal pools, tinajas, pocosins; Figs. 3p–r) and roadside ditches (Fig. 3s) to relatively
permanent vegetation-choked marshes, semi-lotic alluvial swamps, and hardwood depressions
(Figs. 3t–w), some of which are protected parks and wildlife refugia such as the Everglades and

FIG. 3P. Rock outcrop pool in Australia (known also as a tinaja or gnamma). The vegetation growing around the
pool consists of mosses and carnivorous plants, such as sundews.

FIG. 3Q. Aerial photograph of ephemeral playa wetlands in Texas.

FIG. 3R. Playa pool in the Mojave desert.

FIG. 3S. Ephemeral alpine wetland pool in the southern end of the Cascade Mountains in California.

Okefenokee. The origin of wetlands varies tremendously, including, for example, wind-formed depressions (playas), tectonic basins, and alluvial formation.

Because wetlands are generally more ephemeral than other lentic ecosystems, their biotic communities are strongly influenced by how long they retain water (hydroperiod). Wetland invertebrates

FIG. 3T. Roadside ditch in the Great Basin Desert on the California/Nevada border forming an ephemeral pool supporting branchiopod crustaceans and other invertebrates able to complete generations rapidly.

are affected by ecosystem permanency (e.g., time period since last exposure), predictability of drying, and season of drying (if at all), as well as nutrient and light limitations. The nature of the biotic community also reflects the volume and depth of open water, current velocity (for alluvial wetlands connected to the river), and types of predators. For example, the presence of fish completely changes the crustacean fauna of wetlands. Invertebrate species frequenting these ecosystems differ in their adaptations for tolerating or avoiding drought and their period of recruitment. Many species in ephemeral environments are ecological generalists and live in both temporary and permanent aquatic ecosystems.

FIG. 3U. Alluvial swamp in the riverine landscape of the Savannah River in South Carolina.

FIG. 3V. A small portion of the large Cheyenne Bottoms wetlands of southcentral Kansas; this is a major resting and feeding spot for migratory birds.

FIG. 3W. Mud Pond, a sphagnum wetland pond in New Hampshire.

C. Hypersaline Lakes and Pools

Natural saline lakes are common worldwide but generally are smaller in size and abundance than freshwater lakes (Figs. 3x–ab). They are frequently encountered in the more arid prairies and the Great Basin of the western USA and in the provinces of Alberta, Saskatchewan, and British Columbia of western Canada, especially where closed basins promote high salt concentrations. The Great Salt Lake of Utah is the largest saline lake in North America. These lentic ecosystems include saltern lakes, which are high in sodium chloride; lakes with large concentrations of sulfates

FIG. 3X. Pitcher plants (*Sarracenia purpurea*) found in boggy areas; these form an unusual aquatic habitat for small invertebrates while also being a deadly trap for some terrestrial insects.

and borates; and soda lakes characterized by abundant sodium carbonates and bicarbonates. Salinities in the Great Salt Lake have been recorded as high as 200 g/L (vs. an average of 35 g/L in the world's oceans). In addition to high salinities, hypersaline lakes differ in pH, metal concentrations, and alkalinity from more typical inland freshwater lakes.

Saline lakes are extremely rigorous environments for animals and plants. Cyanobacteria and some sedges colonize saline lakes, but the submerged macrophyte flora is relatively sparse. This contributes to a depauperate animal fauna, but the major problem is that few inland water invertebrates can regulate their internal osmotic and ionic concentrations at the level required to live in these high-

FIG. 3Y. Saline Mono Lake in arid eastern California; visible in the upper left (but in front of the mountains) is a small volcanic cone.

salinity environments. For example, the only resident invertebrates in Mono Lake, California, are the brine shrimp, *Artemia monica*, and a member of the brine fly genus *Ephydra*. High densities of brine shrimp in Mono Lake are crucial to the survival of some migratory birds which molt flight feathers while at Mono Lake and depend on brine shrimp as their sole food source while in the lake. As saline lakes become less salty, the diversity and density of their flora and fauna are enhanced. For example, the Waldsea Lake of Saskatchewan is more dilute than Mono Lake and as a result has much higher species diversity (albeit still small). These include various zooplankton groups and benthic groups of insects, one snail species, and several crustaceans.

FIG. 3Z. Calcium carbonate deposits called "tufa" within Mono Lake; these reflect the lakes hypersalinity and chemical composition.

FIG. 3AA. Calcium carbonate soda spring near Cuatro Ciénegas, Mexico.

FIG. 3AB. Acidic salt lake in Western Australia which is teeming with branchiopod crustaceans; this photograph illustrates either the shallow nature of the pool or, less likely, the ability of some scientists to walk on water!

Chapter 4

A Primer on Ecological Relationships among Freshwater Invertebrates

I. INTRODUCTION

Because the breadth and depth of topics in aquatic ecology are far beyond the limited scope of any field guide, our goal here is to introduce some concepts that are especially relevant to invertebrate distribution and abundance in lentic (ephemeral pools, wetlands, ponds, and lakes) and lotic (surface and subterranean creeks and rivers) ecosystems. This chapter focuses on physiological, population, and community ecology. You should read about the key physicochemical features of aquatic habitats in Chapter 3 before tackling this chapter. Both chapters are partially derived from "Chapter 2: An Overview of Inland Aquatic Habitats" in Thorp and Covich (2010). Many more details of aquatic ecology can be gleaned from specialty textbooks on stream, lake, and cave ecology (e.g., Hutchinson 1967, 1993, Wiggins et al. 1980, Wetzel 2001, Dodds 2002, Culver and White 2004, Allan and Castillo 2007, Lampert and Sommer 2007, Thorp et al. 2008, Likens 2009). See Chapters 5–27 of this field guide to learn details on the ecology of individual invertebrate taxa.

II. PHYSIOLOGICAL ECOLOGY

Physiological ecology (or environmental physiology) of inland water invertebrates is the study of how organisms cope with physical (e.g., light, temperature) and chemical (e.g., oxygen, pH, ions) features of the environment through internal biophysical, biochemical, and physiological processes. Interactions with other organisms of the same or different species may be involved, but the primary focus is on the response of an individual in one species.

A. Temperature Relations and Thermal Stress

Temperatures are highly stable in aquatic systems compared to terrestrial habitats, but the amount of temperature fluctuation in aquatic systems over short and long periods varies with habitat characteristics. Mean temperatures generally increase at lower altitudes and latitudes, and the variability is usually greater in shallow than deep waters of both lentic and lotic ecosystems, in open versus shaded areas, and in arid compared with forested ecoregions. Macrohabitats in surface streams within the temperate zone that are affected by ground water inputs are generally cooler in summer and warmer in winter than in surrounding, unaffected areas. Temperatures in streams within caves or flowing

ISBN 978-0-12-381426-5, DOI: 10.1016/B978-0-12-381426-5.00004-1

from them fluctuate less than in other surface streams, but the mean temperature in subterranean streams typically reflects the temperatures of the upstream surface waters averaged over the year. Temperatures in thermal springs are high and constant over all periods. Invertebrates tend to be more tolerant of high temperatures than are fish, but very few species tolerate temperatures above 40°C for long periods.

The timing of emergence, an invertebrate's tolerance of specific habitats, and the annual productivity of individual species and communities are all influenced directly or indirectly by the number of degree days in those habitats (defined by the sum of mean temperatures on each day above a specified minimum value over a relevant life history period), the maximum and minimum temperatures experienced, and how long the organism is exposed to those conditions. Low temperatures influence growth rates, but they rarely directly kill an aquatic invertebrate unless its tissues freeze or ice scours the bottom. High temperatures, however, commonly kill aquatic invertebrates either directly (e.g., destruction of enzymes) or indirectly through reduced oxygen and/or higher energy demands. Tolerance for temperature fluctuations depends on whether the species is stenothermal (narrow temperature tolerance range) or eurythermal (wide range). This can be best judged by laboratory studies of physiological tolerance.

B. Extreme Osmotic and Ionic Conditions

Other than the terrestrially derived aquatic insects (which ultimately had a marine ancestor), most inland water invertebrates evolved directly from estuarine or marine invertebrates (pulmonate snails took a circuitous route that involved secondary colonization from terrestrial habitats). The dilute nature of freshwaters has impeded colonization by many marine taxa, probably to the benefit of aquatic insects. As a consequence of their marine origin, freshwater invertebrates must maintain a different and more concentrated internal ionic and osmotic condition than in the surrounding freshwater. This requires that they be efficient at osmoregulation (almost all species) or at the minimum be able to regulate certain ions.

Animals with either partially calcareous exoskeletons (e.g., crayfish, ostracods, and cladocerans) or hard calcareous shells (snails and mussels) have added needs for calcium ions. This limits their distribution in softwater streams and lakes. Molluscs can live in some softwater streams because they tolerate extremely dilute tissue fluids (the lowest of any metazoan), but they are excluded from some streams in watersheds dominated by igneous rocks rather than limestone. The minimum concentration of Ca^{2+} tolerated by mussels (2–2.5 mg Ca/L) is just adequate to prevent the rate of shell dissolution from exceeding the rate of shell deposition. Despite their heightened need for calcium, crayfish are abundant in many softwater streams of southeastern USA where these ions are difficult to obtain. Decapod crustaceans in general are among the most efficient osmoregulators, but freshwater species face a critical ionic problem when they molt their exoskeleton (Fig. 4a). When external ionic concentrations are low, a crayfish will consume its shed skin to extract additional calcium ions.

At the other end of the osmotic spectrum are hypersaline environments, such as mineral springs and alkaline and brine lakes, where the external environment is more concentrated than the animal's internal fluids. Hypersaline habitats are not only much more concentrated than freshwater and sometimes even seawater (the Great Salt Lake can be five times more concentrated than ocean waters) but they also contain a different ratio of ions. Diversities are universally low in hypersaline systems, and the most concentrated

FIG. 4A. Early molt stages in the crayfish *Procambarus alleni.*

FIG. 4B. The brine shrimp *Artemia franciscana*.

habitats contain only a handful of species at best. For example, the only resident invertebrates in hypersaline Mono Lake, California, are brine shrimp (Fig. 4b) and brine flies.

C. Low Oxygen

Oxygen levels vary greatly among natural and polluted environments, and lower levels can affect the distribution, diversity, and abundance of species. Oxygen levels in streams generally increase with greater turbulence and flow rates and decrease with altitude, temperature, and amount of suspended organic matter. Levels in lakes are similarly affected by many of those factors. In addition, oxygen varies substantially between upper and lower layers of a lake in response to the temperature, amount of bacterial decomposition in the lower layer (hypolimnion), and amount of primary production in the upper layer (epilimnion) (Fig. 4c). Algae and plants produce oxygen during daylight but absorb it at night. When production levels are low (oligotrophic conditions) or moderate (mesotrophic), oxygen levels stay high enough in the epilimnion at all times for invertebrates and fish. However, when production is very high (eutrophic), nighttime oxygen levels may fall to lethal levels for animals; this is aggravated by algal deaths and subsequent bacterial decomposition in the epilimnion. Oxygen levels in lakes and streams are almost always lower in the substrates than in the overlying water column.

It is not unusual for lotic and lentic ecosystems that have been polluted by organic matter to face periodic kills of fish (most often) and invertebrates from low oxygen tensions. Salt lakes and pools and thermal springs are typically low in oxygen because fewer oxygen molecules can be held per milliliter under those conditions. Cold springs and groundwater inputs in general are sometimes low in oxygen because of demands from underground bacterial decomposition and reduced or absent atmospheric recharge from turbulence and normal diffusion.

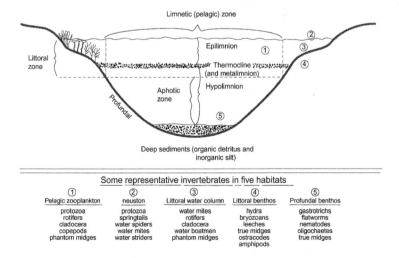

FIG. 4C. Biotic and abiotic zones within a lake along with a list of some representative freshwater invertebrates found within these zones.

FIG. 4D. Ventral view of the pulmonate snail *Physa* crawling on the underside of the water surface.

FIG. 4E. The bloodworm *Chironomus* (Diptera, Chironomidae) which contains a hemoglobin-like respiratory pigment which produces the true midge's characteristic color.

Some species are excluded from environments with regularly low oxygen levels, while other species either tolerate or have strategies for coping with periodically low oxygen tensions. Some species, such as the pulmonate snail *Physa* (Fig. 4d), regularly climb out of the water in small headwater streams or the littoral zone of lakes when oxygen levels drop (this behavior is also used to avoid some aquatic predators). The snail then respires atmospheric oxygen through a heavily vascularized body wall chamber. Many crayfish will move to the shoreline under low oxygen conditions and obtain atmospheric oxygen while still keeping their gills moist. Other species, such as larval midges in the genus *Chironomus* (Fig. 4e), have hemoglobin-like respiratory pigments to enhance uptake of oxygen molecules from the surrounding low oxygen conditions. Some insects living in the substrate, such as dragonfly nymphs in the family Gomphidae, suck water with more oxygen from the overlying water column using an anal tube, while other insects pump water over gills or through protective cases to increase oxygen flow to gills or their general body surface. Finally, many larval flies, bugs, and beetles obtain oxygen directly from the atmosphere or indirectly from the xylem or other tissues of aquatic vascular plants or they carry a bubble of oxygen below the surface as an oxygen reservoir.

III. ORGANISMAL AND POPULATION ECOLOGY

A. Life History Strategies

Life history strategies have been a hot topic in general ecology for a long time, but the focus has rarely been on invertebrates. Our discussion here is limited in breadth and is necessarily confined to freshwater invertebrates, so we recommend you explore the topic more thoroughly in any general ecology textbook.

The life span of most invertebrates is less than a year, though species living in northern latitudes and high altitudes tend to live at least twice as long. Some prominent exceptions to the 1-year or less life span are subterranean species (troglobitic crayfish may live 50 years or more), surface-dwelling crayfish (2–3 years is not unusual even in southern species), larger snails (a couple of years), and unionid mussels (a few years to 200 years in rare cases).

Short-lived species tend to reproduce only a single time (semelparous) and produce many young, whereas long-lived species typically have delayed maturation but reproduce many times (iteroparous).

FIG. 4F. Young crayfish (*Procambarus acutus*) clinging to the abdominal appendages of their mother.

The former are characteristic of r-selected species and the latter of k-selected species (see a general ecology text for an explanation of these two general reproductive/life history strategies). K-selected species tend to be larger in body size and produce fewer but larger young. However, this is not always the case, as shown by freshwater mussels which disperse many small progeny over multiple years. Parental care is more common in k-selected species, but this is relatively rare in freshwater invertebrates. Crayfish are one exception because the young hold on to the female abdomen (Fig. 4f) or forage close by and return when threatened until they are large enough to leave permanently.

Depending on the species and environmental condition, an invertebrate species may produce one (univoltine), two (bivoltine), or multiple generations in a year (multivoltine) or a single generation may require 2 or more years (semivoltine). Multivoltine species, such as chironomid midges, can rapidly exploit new resources and rebound quickly from disturbances such as floods. In contrast, semivoltine species and k-selected species in general tend to be better at resisting disturbances and competing for resources.

B. Dispersal

With the exception of large lakes and rivers (surface and subterranean), all individual inland water ecosystems are relatively ephemeral from the standpoint of evolution of species. Therefore, for a species to survive and adapt to new conditions, they need to either disperse to protected areas of their home aquatic ecosystem when waters evaporate, enter a resistant stage, or find a new ecosystem. Those in intermittent streams may seek shelter in the substrate, move upstream into more permanent headwater springs if present, pass downstream into larger streams, enter a resistant diapause stage, lay eggs (with the adults then dying), or in some limited cases move overland to nearby streams. Invertebrates in ephemeral pools typically enter diapause, lay eggs, or disperse to more permanent pools. The last strategy is accomplished as flying adults, wind-borne propagules, or hitchhikers on the outside of waterfowl or inside the bird gut as resistant eggs, cocoons, or cysts.

Some invertebrates in relatively permanent aquatic habitats lack life history adaptations allowing easy dispersal to new lakes or rivers, while others easily disperse as adults (insects) or by wind as resistant stages when they are stranded on dry land. Travel overland is impossible or at least treacherous for most invertebrates, though possible for short distances for some snails and crayfish, especially in humid environments.

Isolation over thousands of years in a lake, surface stream, or subterranean habitat can lead to evolution of new species. Because of their limited distribution, such new species may be especially vulnerable to extinction from a variety of anthropogenic disturbances including climate change. Moreover, the introduction of competitive or predatory species from other habitats can easily upset the normal community regulatory processes and lead to loss of the native fauna. This is happening with increasing frequency from the pet trade and from other avenues of transport by humans.

C. Adaptations to Currents and Drift

Stream invertebrates live in an environment where downstream displacement by water currents is a nearly constant threat/opportunity. Invertebrates suspended in water currents are part of the stream "drift." Various organisms, such as some mayfly families, are common members of the drift, while many others that are smaller or larger rarely drift. An invertebrate voluntarily entering the drift may do so to escape a predator or aggressive competitor, find better food and space resources, escape a drying stream, or disperse for reproductive reasons. Catastrophic drift is an involuntary phenomenon where the invertebrate is forcibly flushed from a habitat into the current. Whether entering voluntarily or accidentally, drifting exposes the invertebrate to predation from trout, other fish, and even caddisfly nets. It should come as no surprise, therefore, that drifting activity peaks in the evening when visually searching fish are much less effective. This is frequently cited as evidence of behavioral (voluntary) drift, but it may also reflect the fact that invertebrates are more likely to forage in open areas in the evening, such as on the top of algal covered rocks, where they are more susceptible to being lifted into the currents.

Downstream drift would seemingly soon depopulate upstream areas. This tends not to happen because (a) adult stages of insects tend to fly upstream to lay their eggs; (b) aquatic stages of some benthic species regularly move upstream on a daily basis; and (c) survival rates of species in upstream populations would be higher if densities were lowered by drift, thereby balancing the populations and providing enough adults in all areas for reproduction.

D. Vertical and Off-Shore Migration

Most invertebrates spend their entire aquatic stage within a few meters of a central location when not dispersing or accidentally drifting, but some species forage over large areas (e.g., some crayfish) and may even migrate vertically or horizontally on a daily (diel) basis (zooplankton, especially cladocerans and copepods). Vertical migration among lake zooplankton generally follows the pattern of migration from the deep pelagic zone (open water and away from nearshore littoral area) to the shallow pelagic zone shortly after dusk, followed by the reverse movement just before dawn. Shallow pelagic waters have more nutritious algae over the entire 24-hr cycle, but they also have more visually hunting fish predators and greater intensities of damaging ultraviolet radiation (UVR) during daylight hours. In addition to providing a haven from most fish predators and UVR, deep waters are generally colder and thus promote lower metabolic rates which stretch out food resources. Horizontal migration from shallow littoral areas during the day to pelagic surface waters at night also occurs in some taxa. Although the reasons for this horizontal migration are not as clear, they may involve predation (lower in shallow waters among protective aquatic weeds), food resources, and turbulence.

E. Life in Subterranean Ecosystems

Subterranean rivers are most commonly found in karst formations where limestone dissolves over time to produce underground caverns supplied by surface waters. Caves are karsts with an opening to the surface where a stream exits or enters and where surface organisms can gain entry. Organisms adapted to these generally isolated environments experience very constant temperatures, dim or no light, minimal disturbances, and food resources (mostly bacteria-rich, colonized, dissolved organic matter as the food base) which are severely limited in quantity, quality, and abundance.

Obligate cave invertebrates (troglobites) often have special life history adaptions, including exceptionally long lives, delayed maturation, extended reproduction period, fewer but larger eggs, and low metabolic rates. For example, a female crayfish of the troglobitic species *Orconectes inermis* may carry only a few dozen large eggs, while nearby epigean individuals in the same genus can bear hundreds of small eggs. They may also lack pigmentation and functional eyes (see Fig. 3k). Because the character and amount of food are limiting, the food web complexity is also severely reduced to basically detritivores and predators/scavengers.

Densities and species richness (number of species) of aquatic organisms in caves are universally low, but species evenness and endemism (unique species found nowhere else) can be high because of the long-lasting geographic isolation. Karst systems with cave openings are also home to many troglophilic (facultative) aquatic and especially terrestrial species which are commonly found in caves but can also exist in other habitats. These species transport organic matter into the cave (usually as droppings) which eventually enters the karst food web.

IV. COMMUNITY ECOLOGY

A. Food Webs and Feeding Strategies

Aquatic food webs are supported by some combination of autochthonous (internal) and allochthonous (external) sources of primary production. Examples of the former are algae, cyanobacteria, mosses, and aquatic vascular plants, while the latter include organic matter derived from living or decomposed terrestrial plants of all types. These sources vary in nutritive value and ease of consumption. As a general rule, individual algal cells or small colonies are the best source of food at the base of an aquatic food web. Cyanobacteria and many aquatic plants are toxic or at least less palatable or physically difficult to consume. Filamentous macroalgae are also difficult for most invertebrates to eat. Mosses are preyed upon by only a few species of insects. Terrestrial vegetation may be consumed as detritus, with much of the nutritive value coming from the microfloral colonizing it. The food web of heavily canopied streams is often based on terrestrial organic matter, whereas algae typically support the base of the food web in more open areas. Lentic food webs receive most of their nutrition from algae, though aquatic vascular plants contribute to the web mostly as detritus. The kinds of invertebrates common in a given portion of a wetland, lake, or stream reflect in part the diversity and amount of food available for consumption.

FIG. 4G. Aeshnid dragonfly nymph; note the large eyes of this hunting predator.

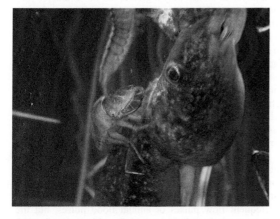

FIG. 4H. Larval water tiger beetle (Coleoptera, Dytiscidae) which has just killed a tadpole.

Feeding relationships within a community vary from very simple and rare food chains to highly complex and common food webs. Although most aquatic invertebrates are feeding generalists or omnivores (eating a diversity of organisms) when you consider their entire life cycles, scientists often group them into different trophic levels, such as producer, herbivore (consumer of algae or plant tissue), detritivore (eating dead organic matter), invertivore (predator on other invertebrates), and higher carnivore/predator levels. Scavenging on recently dead and mostly intact prey can be considered predation. Consumers can also be classified according to how they obtain their food, such as algal scrapers, plant/algal piercers, shredders, collector/gatherers, collector/filterers, suspension feeders, sit-and-wait and hunting predators (Fig. 4h), parasites and parasitoids, detritivores, and omnivores. Most prey of invertebrates are other invertebrates, but some large and voracious invertebrates, such as water tiger beetles, can kill small fish and amphibians (Fig. 4h). Some categories overlap, such as suspension feeders (e.g., a mussel that eats living and dead organic matter in suspension in the water column) and planktivore (an organism that eats phytoplankton or zooplankton).

Some food webs are simple enough to be considered food chains, but most are very complex. Two examples of the former are (a) algae, brine shrimp, and waterfowl in saline Mono Lake; and, (b) dissolved organic matter, bacteria/fungi, invertebrate detritivore, and predator in subterranean rivers. Complex food webs are ones with the most trophic levels, species, and pathways among species and trophic levels.

Food web relationships can be studied through behavioral observations (who eats who), gut content analysis, and chemical analysis. The first is difficult to observe in most species, while the second provides information on what is ingested rather than assimilated and is limited to items consumed over perhaps 24 hr. Chemical analyses, such as stable isotope and lipid analyses, provide a longer term feeding picture but are less exact about the species consumed.

B. Top-Down and Bottom-Up Control of Food Webs

To varying extent, the abundance, diversity, and distribution of aquatic invertebrates are controlled by some combination of competition for food and space, predators, parasites, and physicochemical features of the habitat. Competition may involve the ability to control space through overt or threatened aggressive interactions (interference competition) or the ability to harvest resources faster or more efficiently than a competitor (exploitative competition). Where food is insufficient, "bottom-up" factors are deemed most important for the species or community. The availability of algae might limit the production of benthic invertebrates in streams where the canopy cover is heavy. In contrast, where predation regulates diversity and density, the system is considered to be controlled by "top down" factors. Keystone species are predators that exert a dominant effect on the diversity of the overall community. In food chains and relatively simple food webs, a "trophic cascade" may exist where a top predator reduces the density of the level below it, which causes a sequence in lower levels of increases followed by decreases, and so on. Parasitism is also an example of top-down control, but its importance has rarely been evaluated for aquatic invertebrates. Scientists also know almost nothing about the role of viruses in affecting populations and communities of invertebrates, although there is some evidence that viruses may reduce bacterial production by half, which in turn would eventually affect availability of food for the invertebrate community.

C. Physical Control of Communities

While interspecific competition and predation are present in all aquatic ecosystems to some extent, physical factors also play a role in determining the distribution and abundance of many invertebrates. The predominant view among ecologists is that physical factors play a greater but not exclusive role in controlling diversity within lotic than lentic systems. However, this consensus tends to ignore ephemeral pools, where the length of the hydroperiod is critical to ecosystem structure and function. Within permanent systems, however, streams are probably more subject to physical disturbances than lakes. Small streams and those in more arid regions experience more variable and less-predictable pulses of water (called flood pulses if the stage height exceeds bankfull) following rain events than do larger lotic systems. As flow pulses increase in severity, more of the substrate is moved, thereby disturbing the resident invertebrates and their food resources.

While flow pulses may reduce overall densities in a stream by disturbing substrates and washing species downstream or on to land, this physical disturbance also favors species able to reproduce rapidly or seek shelter from the high currents. In this way, the physical disturbance may prevent a superior competitor from driving a subordinate to extinction and thereby allow greater species diversity in the community. Flood pulses also benefit species that need to reproduce on the flooded floodplains. Fish that are floodplain specialists especially benefit from regular, predictable, and extended flood pulses, but this may also favor many invertebrates, though this has rarely been studied. Stream ecologists now believe that a natural flow regime of fluctuating flows is vital to the health of all lotic ecosystems.

Ecologists often focus on the effects of frequent flow pulses or annual flood pulses, but the periodicity and degree of floods and droughts over periods of 1–100 years (flow history) and >100 years (flow regime) can be very important to aquatic and floodplain species of animals and plants.

Another physical disturbance is ice. Far northern wetlands frequently freeze to the bottom, but resident invertebrates typically have life history adaptations allowing their survival. Streams less commonly freeze to the bottom, but whenever ice touches the bottom of a stream or wind-swept littoral zone of a lake, it can scour the bottom and eliminate all surface species. Stream communities will rapidly recolonize after the thaw, but shallow lake bottoms may permanently show a difference in community composition within and below ice-scoured areas.

All benthic invertebrates are strongly affected by the size, texture, permeability, stability, composition, and location of the substrate. Substrates provide sites for resting, food acquisition, reproduction, and development as well as refuges from predators and inhospitable physical conditions.

D. Invasive Species

Native species have evolved and adapted to their normal habitats and neighbors over many thousands of years, but the presence of individual species and the structure of the community can be severely disrupted in a matter of a few years by the invasion of a species foreign to that ecosystem or even continent. Expansion and contraction of species ranges are a matter of history, but the rate and degree of change has accelerated tremendously from the intentional and accidental effects of humans. Some

FIG. 4I. The invasive bivalve mollusc *Dreissena polymorpha*, which first invaded North America through the Great Lakes but which has now spread to the majority of states.

FIG. 4J. The Australian yabbie, *Cherax quadricarinatus*, is an invasive species which should soon reach the southwestern USA from Mexico where it escaped from aquaculture facilities.

truly exotic species, such as the zebra mussel, *Dreissena polymorpha* (Fig. 4i), have arrived accidentally from Europe in the fresh ballast water of seagoing ships. Other species, such as the Australian yabbie, *Cherax quadricarinatus* (Fig. 4j), which reaches multiple kilograms in weight, were intentionally introduced for aquaculture or escaped from aquaculture facilities. International aquatic species also arrive through the pet trade and are then released by ill-informed aquaculturists who may be reluctant to kill the imported animal or plant and thus dump it into a local pond or stream when they tire of their hobby. Some native species, such as the red swamp crayfish, *Procambarus clarkii* (Fig. 4k), have been intentionally spread through much of the USA and several foreign countries for bait or human food production. Species native to North America also exploit new opportunities for dispersal when formerly isolated water bodies are linked by canals, such as the canal now joining Lake Michigan and the Mississippi River, or when human disturbances to river channels or watersheds substantially change the aquatic habitat and thereby favor non-native species.

Scientists have not been terribly successful in predicting the fate of an actual or potential invasive species. If a niche is open, of course, an invader is more likely to be successful. This

FIG. 4K. The red swamp crayfish, *Procambarus clarkii*, which is raised for aquaculture throughout much of the southeastern USA and has unfortunately been exported to some European and Asian countries.

happened with zebra and quagga (*Dreisenna bugensis*) mussels. In many cases, however, there did not appear to be an open niche, and success may have depended on superior ability to compete, avoid predation, reproduce faster, and/or tolerate adverse environmental fluctuations.

To preserve our native species of aquatic invertebrates, it is crucial that we do a better job on large and small scales in preventing the accidental or intentional introduction on non-native species.

Ecology and Identification of Specific Taxa

Chapter 5

Sponges: Phylum Porifera

I. INTRODUCTION, DIVERSITY, AND DISTRIBUTION

Freshwater sponges are a moderately common, occasionally abundant, and frequently overlooked component of the benthic fauna of creeks, rivers, ponds, and lakes (Fig. 5a). They may have encrusting, rounded, or finger-like growth forms. Sponges are often mistaken for algae because many are green from the large internal flora of symbiotic algae which provide nutrition for the sponge in exchange for a protected habitat within the sponge and other benefits. These simple animals can be important filter feeders in some ecosystems that are relatively unpolluted and free from heavy suspended sediments.

Scientists presently lack a clear consensus on the "individual" versus "colonial" nature of sponges, in part because of the ongoing debate on what constitutes an individual organism in nature. In some ways sponges function as colonies, but their very simple, tissue level of organization— perhaps the least complex of all multicellular organisms—makes it difficult to compare their status with other, more clearly colonial animals, such as marine corals (phylum Cnidaria).

The 30 or so North American freshwater sponge species are moderately small, though giant colonies have been found spread over many square meters and with an estimated $1\,m^3$ volume. This contrasts with the more diverse marine species in the phylum Porifera which range in size from small sponges that bore into live oyster shells to giant basket sponges large enough to hold a human diver.

All freshwater sponges are members of the class Demospongiae (order Haplosclerida, suborder Spongillina), the most morphologically complex and species rich class in the phylum. Most North American species are in the cosmopolitan family Spongillidae, with classification based primarily on the structure of one of three major forms of microscopic silica spicules in the sponge skeleton. Because these support structures require a high-powered light microscope for identification, our classification in this field guide is restricted to this family level.

II. FORM AND FUNCTION

A. Anatomy and Physiology

From a macroscopic perspective, freshwater sponges can develop an encrusting, rounded, or finger-like growth form, all within the same species but usually under different habitat conditions of light and flow rate. Internally, however, sponges are anatomically simple organisms with tissues as their highest level of organization. Their basic organization consists of an outer, epidermal layer covering an inner organic matrix, or mesohyl. A water canal system is interspersed throughout this structure. Inorganic siliceous spicules bound together by an organic collagen matrix support the sponge body.

Specialized sponge cells accomplish many basic biological functions that are handled by organs in more complex phyla, including structural support, respiration, food uptake and digestion, excretion,

ISBN 978-0-12-381426-5, DOI: 10.1016/B978-0-12-381426-5.00005-3

FIG. 5A. A colony of the freshwater sponge *Spongilla lucustris* showing digitform growth.

and reproduction. The water canal system of progressively finer filtering chambers and passageways transports vital food and oxygen, disposes of waste products, and serves as a conduit for reproductive propagules. The filters and specialized cells within the canals and mesohyl can extract particles from the passing water ranging in size from large algae to small bacteria and perhaps even larger dissolved organic matter.

B. Reproduction and Life History

A sponge's life cycle includes periods of dormancy, active growth, and both sexual and asexual reproduction. Asexual reproduction occurs via physical fragmentation of the "adult" sponge and through production of resistant, multicellular gemmules, which later hatch to renew the active growth period. Most sponge individuals appear to be unisexual at any given time but can change gender over time.

III. ECOLOGY, BEHAVIOR, AND ENVIRONMENTAL BIOLOGY

Sponges are filter feeders and hosts for symbiotic algae (a relatively uncommon relationship in freshwater taxa). They can filter substantial numbers of bacteria and suspended algae from the water, making them serious competitors with some protozoa, zooplankton, and a few other multicellular taxa. A single finger-sized sponge can apparently filter nearly 125 L of water (~33 gal) in a day. Aside from the filtration impact, the sponge's algal symbionts contribute to benthic primary productivity in lake and stream ecosystems. Contrasting with their role as important consumers of plankton, their silica "skeleton" and various toxic and pharmacologically active compounds make them relatively invulnerable to predators. Consequently, their direct energy contribution to higher aquatic food webs is essentially blocked, and mostly occurs indirectly through detrital pathways. Only a few insect species prey on sponges, such as spongilla flies (order Neuroptera, which includes the terrestrial antlions and their relatives). Sponges also function as physical habitat for a variety of largely innocuous invertebrates that live on or in the sponge body. These include worms, small clams, mites, and various insects.

Many but not all sponges have evolved a symbiotic relationship with algae (typically green algae but at least in one case with a yellow-green algal species). This relationship ranges from facultative (optional) to obligate and seems to vary with prevailing environmental conditions. In such symbiotic relationships, the host provides shelter, protection from predators, and a source of nutrients (nitrogen and phosphorus) and carbon dioxide for photosynthesis. In return, the symbiont provides photosynthetically-fixed carbon to the invertebrate host. These algal symbionts are known to contribute as much as 50–80% of the energy needed for sponge growth, based on a pond study in the northeastern USA.

Many sponges have a rather cosmopolitan distribution, but almost all require (i) a hard surface on which to grow; (ii) an environment that is not overly disturbed by physical fluctuations; and (iii) water that is neither too polluted nor loaded with large amounts of silt, sand, or other particles that could clog their filtering system. Light availability can also be a limiting factor for species with symbiotic algae.

IV. COLLECTION AND CULTURING

Sponges can be collected from shallow water by hand or with a long-handled rake or even by snorkeling in shallow water. Look for them on the top of cobble, among rooted aquatic plants, or on a submerged log or tree limb. In addition, many species without symbiotic algae can be found under stones and on the undersurface of overhanging banks.

Growing sponges in an aquarium is too demanding for all but experienced scientists because of the demands of water quality and adequate supplies of living, microscopic food sources.

V. REPRESENTATIVE TAXA OF FRESHWATER SPONGES: PHYLUM PORIFERA

A. Class Demospongiae, Family Spongillidae

Freshwater Sponges (Fig. 5b): *Anheteromeyenia* sp., *Corvomeyenia* sp., *Corvospongilla* sp., *Dosilia* sp., *Duosclera* sp., *Ephydatia* sp., *Eunapius* sp., *Heteromeyenia* sp., *Racekiela* sp., *Radiospongilla* sp., *Spongilla* sp., *Stratospongilla* sp., and *Trochospongilla* sp.

Identification: Brown, yellow to gray and white or green, forming crusts on rocks, branches, or plants, or rarely developing as stalked structures. **Size:** Forming mats that cover areas of a few square millimeters to several square yards. Typically start out as a semicircular crust that spreads outward. **Range:** Subarctic regions of the world; common throughout North America. **Habitat:** Permanent lakes and ponds, rivers, and streams. Typically found on the underside of rocks or overhanging banks. Green forms grow on top of rocks and on wood or twigs. **Remarks:** Often mistaken for algae. These colonial organisms are fed upon by the larvae of spongillaflies (see Chapter 24).

Freshwater Sponges (Fig. 5c): *Digitform growths, with symbiotic algae.*

Freshwater Sponges (Fig. 5d): *Encrusting growth form, with gemmules (black dots). A case with four light-colored snail eggs in on the left side.*

Chapter 6

Hydra and Jellyfish: Phylum Cnidaria

I. INTRODUCTION, DIVERSITY, AND DISTRIBUTION

The common brown and green hydra and the rarer freshwater jellyfish *Craspedacusta sowerbii* are anatomically simple, fragile animals found widely throughout North America, south of the Arctic. They are closely related to the hydroid *Cordylophora caspia*, which lives in large deep rivers where it forms sparsely branching colonies up to 5 cm high. *Cordylophora* is native to the Caspian and Black Seas of Eurasia and may have reached North America via ship ballast waters and then spread widely by attaching to commercial and recreational boat hulls. Other North American freshwater cnidarians include the microscopic colonial hydroid *Calposoma* and *Polypodium hydriforme*, which begins life as a tiny parasite on fish eggs. After the fish eggs hatch, *Polypodium* develop into adults that use their tentacles to "walk" along the bottom, where they hunt worms. These animals are only a millimeter in length and so are very difficult to find.

Scientists consider freshwater cnidarians to be evolutionarily ancient members of the mostly marine class Hydrozoa. They are distantly related to other marine cnidarians, such as corals, sea anemones, and true jellyfish.

II. FORM AND FUNCTION

A. Anatomy and Physiology

All freshwater cnidarians have a radially symmetrical, two-cell layer, tubular body separated by a thin noncellular layer called the mesoglea. Their combined mouth–anus structure is ringed with outwardly projecting tentacles containing stinging cells (nematocysts). Stings from a few hydra are not noticeable to most people; but when populations reach the millions on fishing nets left submerged for long periods, the combined stings have been known to cause rashes on the hands of fishermen. Hydra exist only as polyps, whereas *Craspedacusta* spends part of its life as a dome-shaped, free-swimming medusa and the remainder as a benthic polyp. Cnidarians lack eyes and internal organs.

B. Reproduction and Life History

Hydra may sexually produce eggs or have offspring that bud off asexually from the parent, but asexual budding occurs more frequently than sexual reproduction. Simultaneous or sequential hermaphroditism (both sexes present at some time) is the rule in hydra. Colonies can form by incomplete separate of an asexual bud in the polyps of all freshwater species except hydra.

Freshwater jellyfish reproduce by eggs that develop into a microscopic hydra-like animal; this stage then forms branches which produce buds that develop into tiny individual jellyfish. With the exception of an African species, freshwater jellyfish medusae do not produce other medusae by asexual budding.

ISBN 978-0-12-381426-5, DOI: 10.1016/B978-0-12-381426-5.00006-5

III. ECOLOGY, BEHAVIOR, AND ENVIRONMENTAL BIOLOGY

Freshwater cnidarians are benthic animals, with the exception of the medusoid stage of *Craspedacusta* and periodically dispersing and floating hydra. They attach to firm surfaces, including aquatic plants, but avoid surfaces with a heavy biofilm layer (bacteria, algae, etc. in an organic and inorganic matrix). To move between habitats in search of more plentiful prey, jellyfish float while the mostly sedentary hydra either walk with a somersaulting motion or float with a bubble.

Freshwater hydrozoans are predominately predators, though green hydra also derive nutrition from internal, symbiotic algae. Hydra use stinging cells to subdue and then consume small crustaceans, worms, mosquito and other insect larvae, and even occasionally small larval fish. In turn, they are preyed upon by flatworms, crayfish, large amoeba such as *Hydramoeba hydroxena*, and probably other predators that can tolerate the stinging cells.

Hydra live in waters at temperatures from nearly freezing to around 25°C and occur so widely in this and many other continents that the chances of extinction of a whole species are considered minimal.

IV. COLLECTION AND CULTURING

Hydra can be found in reasonably unpolluted waters where they attach to rocks or aquatic plants in many slow stream and lake shore habitats. They tend to be most abundant above the outfall of lakes or in a calm stream pool below the lake where zooplankton prey are abundant. To find hydra, collect cobble, decaying terrestrial leaves, or fresh plant material from these aquatic habitats and place them in a glass or enamel (white preferably) pan with the cool lake or stream water and allow this material to settle overnight. Later carefully examine the sides of the pan and upper surfaces of plant matter and rocks for hydra using a magnifying glass. Be careful not to jostle the pan because the hydra will contract, making them harder to see.

Unfortunately, the odds of finding freshwater jellyfish are like winning a moderate-size lottery prize. Their appearance in a given lake or even the same lake but in different years is very difficult to predict. If spotted, however, they can be collected gently with a bucket and very carefully transferred to an aquarium containing lake water at a similar temperature.

Hydra and jellyfish can be kept in aquaria and fed live zooplankton or brine shrimp larvae. Because of the fragile nature of jellyfish, the aquarium should be as large as possible, with little or no disturbance to the water. Jellyfish are very sensitive to changes in dissolved oxygen, so should be kept as cool as possible to prevent suffocation. If food is plentiful, an aquarium can become overrun by multiple generations of hydra, while jellyfish are unlikely to reproduce or live long.

V. REPRESENTATIVE TAXA: PHYLUM CNIDARIA

A. The Hydras: Family Hydridae

Brown Hydra (Fig. 6a): *Hydra sp.*

Identification: Brown, yellow to gray and white or translucent hydra. Tentacles may be a third to twice as long as the body. In long tentacled forms the tentacles may be extremely thin. **Size:** 1–25 mm. **Range:** Subarctic regions of the world; common throughout North America. **Habitat:** Permanent lakes and ponds, especially near outflows and on dams; found at depths of 60 m or more. **Remarks:** Brown hydra are occasionally found at incredibly high densities on fishing nets, causing rashes on the hands and arms of the fishermen.

Green Hydra (Fig. 6b): *Hydra viridis*

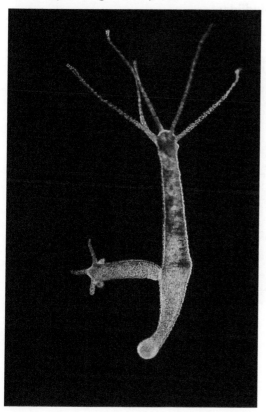

Identification: Green opaque to translucent hydra. Tentacles up to one half the body length. **Size:** 1–20 mm. **Range:** Subarctic regions of the world, common throughout North America. **Habitat:** Permanent lakes and ponds, especially near out flows and on dams, usually where light penetrates; usually within depths of 2 m. **Remarks:** Green hydra are green due to the presence of symbiotic unicellular algae (*Chlorella*), which produce sugars that help feed the hydra. As a result, green hydra can survive for months without animal prey, whereas brown hydra can only last a few weeks.

B. The Clavid Hydrozoans: Family Clavidae

Freshwater Hydroid (Fig. 6c): Cordylophora caspia

Identification: Long, thin, branching colonies affixed to a substrate. Tentacles projecting irregularly at branch apices (compare with bryozoans). **Size:** Colonies grow to 500 mm long. **Range:** Native to Ponto-Caspian region, introduced to Great Lakes, Atlantic coast, Upper Mississippi River system, and large Pacific drainage rivers. **Habitat:** Brackish and polluted waters primarily, but sometimes in unpolluted situations. Typically in large, deep rivers and lakes. **Remarks:** Easily mistaken for moss animals (bryozoans). Forms colonies on any hard substrate, including boat hulls or floats, which can easily be transferred to other water bodies.

C. Hydrozoan Jellyfish: Family Olindiidae

Freshwater Jellyfish (Fig. 6d): Craspedacusta sowerbii

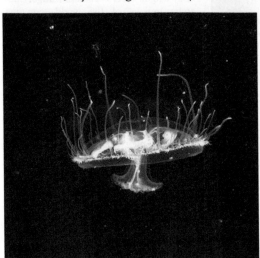

Identification: Circular, convex, bell shaped, with a fringe of tentacles. Under the "bell" is a small central projection called a manubrium. Animal is translucent, and appears to be internally divided into four "pie piece" sections. **Size:** 1–20 mm. **Range:** Subarctic North America, appears to be distributed in freshwater worldwide. **Habitat:** Permanent ponds and lakes. **Remarks:** The only recorded freshwater jellyfish in North America. Occasionally rests on the bottom, typically swims toward surface by contractions of the bell-shaped body then floats down toward the bottom.

Chapter 7

Flatworms: Phylum Platyhelminthes, Class Turbellaria

I. INTRODUCTION, DIVERSITY, AND DISTRIBUTION

Except for two to three species of ribbon worms (phylum Nemertea), all 200+ species of freshwater flatworms in North America are in the phylum Platyhelminthes. The internal parasitic tapeworms (class Cestoda) and flukes (class Trematoda) are also in this phylum, but only the latter has a short-lived, free-living stage where larvae seek new hosts (these cause the temporary irritation known as swimmer's itch). For the purpose of this chapter, the name "flatworm" refers only to turbellarians.

Flatworms consist of almost entirely free-living microturbellarians (a few millimeters or smaller) and macroturbellarians (closer to a 1 cm or more in length in North America), though this is a nonphylogenetic classification. Two common names for freshwater macroturbellarians are triclads and planarians. These include *Dugesia* and *Planaria*, which are widely used in high school and college science classes to study simple behavior and tissue regeneration. Roughly 20% of the North American flatworm species are planarians, with the remainder being microturbellarians.

Turbellarians live in a great diversity of habitats from permanent ponds, lakes, streams, and rivers to temporary pools and wetlands. Some even thrive within semiaquatic habitats such as among wet mosses at the edge of a stream, or within the film of water coating fallen leaves on a moist forest floor, or even in the capillary water within soils of grassy meadows. Several terrestrial species have been introduced to North America, including *Bipalium kewense*, which may grow to 30 cm in length. Flatworms, especially triclads, are important fauna in cave streams and in the upper portions of springs. They have been collected as far north as Alaska and in the Hudson Bay region of northern Canada and are also abundant in tropical regions. Shallow waters of lakes and streams support the greatest density and diversity of flatworms, but turbellarians reach the deepest parts of lakes as long as oxygen is sufficient.

Most flatworms are benthic, though some species can swim using cilia and the steering action of muscular body waves. They commonly rest or hunt in protected areas (e.g., under a stone) during the day and come out when it is dark. Triclads seem more adverse to light than are the microturbellarians. Some vernal pool species contain symbiotic algae, and so spend most of their time in the sunniest portions of the pool. Flatworms typically glide across the bottom by employing cilia for propulsion through a thin layer of mucus they have laid down. Other species may creep along with leech-like movements. Although many species live on the surface of the stream or lake bottoms, others— especially microturbellarians—reside interstitially *within* the substrate where they move among the minute sand grains. Swimming and gliding in the turbellarians is powered by cilia of the epidermis, and the animals steer these movements by musculature of the body.

ISBN 978-0-12-381426-5, DOI: 10.1016/B978-0-12-381426-5.00007-7

II. FORM AND FUNCTION

A. Anatomy and Physiology

Scientists rank turbellarians as the most primitive bilaterally symmetrical, multicellular organisms on earth, in part because of their incomplete gut (only one opening) and lack of the major body cavity (coelom) found in higher organisms. These worms are unsegmented and have three cell layers (ectoderm, mesoderm, and endoderm) in contrast to sponges and cnidarians. All species lack a hard skeleton.

Flatworms have an elongated shape that varies in cross section from flat (especially macroturbellarians) to cylindrical, and some asexually reproducing forms consist of chain-like, cylindrical units (zooids). Their color varies from nearly colorless to white or shades of red, yellow, blue, green, brown, black, or any combination of these colors, depending on pigments in the outer body layer, whether they have symbiotic algae in the epidermis, and even the color of their gut contents. They may appear solid in color or be mottled or striped. The body tends to be tapered at both ends and may have lateral flaps in the anterior cerebral region giving the appearance of a distinct head. Some species have a short tail. Hair-like cilia cover the entire body surface and contribute to various body functions, including locomotion.

Platyhelminths use a variety of sensory structures to obtain environmental stimuli related to touch, pressure, chemical conditions, and light. These sensory receptors may be scattered across the body surface or organized into sensory complexes, such as eyes, chemosensory pits, and statocysts (organelles sensitive to body orientation). One or more pairs of primitive, pigment-cup eyes (ocelli) are present anteriorly in many species. These ocelli present in the familiar laboratory *Dugesia* make the flatworm appear cross-eyed. However, this is misleading because the flatworm eye-cup merely detects light intensity and cannot form images.

Uptake and release of gases occur across the body surface because no discrete respiratory organs are present. Their dependence on body surface exchange is inefficient and thus tends to bar most flatworms from areas with low oxygen. Some species with hemoglobin blood pigments or symbiotic algae (which produce oxygen) can survive low or zero oxygen conditions.

B. Reproduction and Life History

Some flatworms, especially those living in temporary aquatic habitats, reproduce only once per year (univoltine), while other species are multivoltine, with the number of generations depending on environmental favorability. Most flatworms reproduce sexually, and almost all species are hermaphroditic. Although gamete exchange is most common, self-fertilization is possible in some species. Eggs are laid singly on a hard substrate, are grouped into clusters (sometimes stalked), or are placed in cocoons. Miniature flatworms emerge from eggs which have been attached to the substrate. These usually develop directly (without intermediate larval stages) into adults and differ only in size and presence of reproductive organs. A few species are ovoviviparous (young brooded inside parent) or have a distinct larval stage. Asexual reproduction occurs seasonally in many taxa and is the only form of reproduction in a few species. This form of reproduction involves differentiation of the adult body into distinct units (zooids) followed by separation into small flatworms via transverse division of the adult body. This process is aided by the remarkable regenerative abilities of flatworms.

Under demanding environmental conditions, flatworms produce resistant eggs able to survive desiccation of the surrounding environment or the animal may enter diapause (a very low metabolic state) and encyst as a whole animal or fragment of the adult.

Sexually reproducing flatworms can live from a few weeks to a few months (not counting periods of encystment), while the life span of asexual species is theoretically indefinite.

III. ECOLOGY, BEHAVIOR, AND ENVIRONMENTAL BIOLOGY

All triclads and most microturbellarians are predators, and many species scavenge partially on dead organisms. Triclads capture the largest prey types, including various small crustaceans, worms, many insects, and the eggs of fish and salamanders. Microturbellarians have a diverse diet of bacteria, algae, protozoans, and small invertebrates. Flatworms detect prey chemically, by touch, or through vibrations in the water. They then attack the prey and grasp it with their pharynx, a tubular structure that can be protruded to suck up whole prey or sometimes bite off pieces. Solid wastes are later ejected through the pharynx. Prey are sometimes immobilized with poisonous secretions from subepidermal glands, with these secretions also serving to dissuade other predators from capturing the flatworm itself. Other flatworms use mucus to trap prey or merely grasp the prey with their muscular pharynx. Some species hunt prey from the stream or lake bottom, while others float near the water surface on a mucous filament and await planktonic prey or small drifting insects.

A variety of predators feed on flatworms, and they are also attacked by parasites, such as protozoans and nematodes. The predators include fish, predaceous midge larvae, beetle larvae, and other turbellarians.

IV. COLLECTION AND CULTURING

Several techniques are useful for collecting triclad flatworms. A trap baited with fresh liver or an injured insect will attract some species. A simpler way, however, is to collect organic matter from the aquatic system and place it in a tray with clean, nonchlorinated water. You can spot the planarians as they later move over the pan surface. Alternatively, if the organic matter is placed in a gallon jar, the planarians will begin swimming toward the surface as the water begins to stagnate and they seek more favorable oxygen conditions. You can also filter large quantities of water from a pond or stream littoral zone or from a cave through a small mesh net. Finally, if you turn over cobble or larger rocks in a stream, triclads can often be found moving over the surface. [Note: Be sure to return the rock to its former position and orientation to protect the species on it.] The tip of an artist's soft paintbrush is useful in picking up and transferring planarians to a separate container without damaging them.

Large and small flatworms can be maintained indoors in small glass containers or in aquaria as long as the water is kept clean, well oxygenated, and around a normal room temperature of 17–25°C. They should be kept in clear unpolluted water or even synthetic water (distilled water to which a specific mixture of vital chemicals has been added). Food requirements are species-specific, and you will need to try several options, including whole or parts of various small aquatic invertebrates such as water fleas (Cladocera), small aquatic worms, and mosquito larvae—but do not allow the mosquitoes to emerge as adults!

V. REPRESENTATIVE TAXA OF FLATWORMS: CLASS TURBELLARIA

Planaria (Fig. 7a): Planaria sp., Dugesia sp., and Phagocata sp.

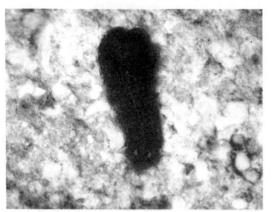

Identification: Black, brown, gray, or white flatworms with an angular head and eye spots. **Size:** 1–10 mm. **Range:** Subarctic regions of the world; common throughout North America. **Habitat:** Nocturnal. Normally found under rocks or on wood in permanent lakes, ponds, rivers, streams, springs, and seeps. Some cave species. Typically found to a depth of 30 m. **Remarks:** Sometimes mistaken for leeches (see Chapter 11). Can sometimes be collected in large numbers using pieces of meat.

Vernal Pool Flatworms (Fig. 7b): Turbellaria sp.

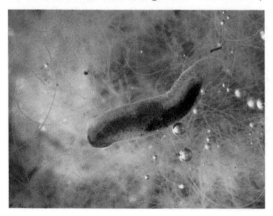

Identification: Black, brown, gray, mottled brown, white, or green flatworms with a rounded head. Eye spots present or not. **Size:** 1–4 mm. **Range:** Vernal pools and other seasonal wetlands in California. **Habitat:** Found during the winter and spring months in temporary wetlands moving over the bottom or plants or cruising along the surface. **Remarks:** Sometimes mistaken for leeches. Often found swarming around injured or drowning insects or crustaceans. Green species have algae living in their bodies which produce food for them.

Chapter 8

Hairworms: Phylum Nematomorpha

I. INTRODUCTION, DIVERSITY, AND DISTRIBUTION

"Hairworms" or sometimes "horsehair worms" of the phylum Nematomorpha are an unusual group of freshwater invertebrates because adults are free-living but larvae are internal parasites. The name reflects an erroneous belief that these worms arose from horse hairs that fell into water—a belief that was disproved in the late nineteenth century.

With the exception of one marine genus, the adults of all species of hairworms live in freshwater, with 20 species reported from North America. Adults occur throughout the world north of Antarctica in many aquatic habitats ranging from watering troughs to the Great Lakes and from shallow streams to large rivers.

Hairworms seem to have no close relative among other invertebrate phyla but have been linked to nematodes, rotifers, and gastrotrichs.

II. FORM AND FUNCTION

A. Anatomy and Physiology

Adult hairworms are long and slender (0.25–3 mm in diameter), with most species reaching several centimeters but some giant worms obtaining lengths of nearly a meter. The most common colors are yellowish brown to black, but occasionally white hairworms are collected.

The body wall of a hairworm consists of an outer cuticle and longitudinal muscles. A lack of circular muscles produces a whip-like action when the hairworm moves. In contrast to the turbellarians (Chapter 7), hairworms have a complete gut with separate openings for the mouth and anus. However, adults do not feed, so these structures and the excretory organs are nonfunctional and typically degenerate.

Hairworms are in a group of phyla pseudocoelomates because they lack the peritoneum-lined body cavity (coelom) of higher animals. They are anatomically more advanced than acoelomate animals (e.g., flatworms), which have a solid body lacking any major body cavity.

B. Reproduction and Life History

The four-stage life history of hairworms consists of the egg, a pre-parasitic larva that hatches from the egg, the parasitic larva that develops within an invertebrate host, and the free-living adult inhabiting freshwater. The initial host (paratenic host) of the pre-parasitic larva may serve only as a temporary

ISBN 978-0-12-381426-5, DOI: 10.1016/B978-0-12-381426-5.00008-9

FIG. 8A. Hairworm emerging from a cockroach host.

carrier for transporting the larva to the larger, final host. This latter "definitive" host usually picks up the parasitic larva following the definitive host's predation or scavenging on the paratenic host. Paratenic hosts are usually invertebrates but can also be vertebrates, especially amphibians or fish. The life cycles of most North American hairworms species probably involve a paratenic host, while the definitive host of most species on this continent are usually relatively large omnivorous insects in the order Orthoptera (e.g., crickets) (Fig. 8a) or predaceous carabid beetles (order Coleoptera). Adults of these large insects seek freshwaters, probably from the parasite's influence, and thereby allow the developing adult horsehair worm to escape, mate, oviposit, and produce the infective larvae which will begin the cycle anew.

Adults have separate sexes and mature rapidly after emerging from their hosts. After mating, the eggs are deposited singly or in clusters on rocks, twigs, or submerged debris.

III. ECOLOGY, BEHAVIOR, AND ENVIRONMENTAL BIOLOGY

The ecology of adult freshwater hairworms is rather simple because the adults do not eat prey and their free-living lives are relatively short. It is not surprising, therefore, that the ecology of the adults has rarely been studied and most relevant information is anecdotal. Because they are relatively large and do not bury in the substrate, hairworms are susceptible to predation from crayfish, a large diversity of fish, and even some terrestrial invertivores such as birds.

IV. COLLECTION AND CULTURING

Look for hairworms in slowly moving streams away from the main current or in pools or ponds. During spring or late summer, dig a small hole in gentle current and wait to see if the adult worms drift downstream into it. You can also collect them with a net if you spot one drifting downstream. Be advised, however, that it may take quite a bit of effort to find these nematomorphs.

Because the life cycle requires a parasitic larval stage and the adults do not eat, the free-living adults cannot be kept alive indefinitely in an aquarium.

V. REPRESENTATIVE TAXA OF HAIRWORMS: PHYLUM NEMATOMORPHA

Hairworms (Fig. 8b): Gordius sp., Paragordius sp., Parachorodes sp., Gordonius sp., Neochorododes sp., Chorododes sp., and other genera

Identification: Black, brown, gray, or tan thin, whip-like worms, sometimes with a bilobed or trilobed end. Eyes lacking. **Size:** 4 cm to more than 1 m long. **Range:** Subarctic regions of the world; common throughout North America. **Habitat:** Pools, puddles, ponds, watering troughs, quiet backwaters along streams and rivers. Typically in shallow water, rarely found below 1 m deep. **Remarks:** Sometimes mistaken for leeches (see Chapter 11) or thought to bite. Adults do not have a functional mouth and cannot bite or feed.

Hairworms (Fig. 8c): Adult hairworm emerging from a carabid beetle host.

V. REPRESENTATIVE TAXA OF HAIRWORMS: PHYLUM NEMATOMORPHA

Hairworms (Fig. 8b). Cotton... sp. : Paragordius sp. (variline)...

Hairworms (Fig. 8c). Adult hairworm emerging from a cricket (male)...

Chapter 9

Snails: Phylum Mollusca, Class Gastropoda

I. INTRODUCTION, DIVERSITY, AND DISTRIBUTION

A. General Features

Ponds, lakes, creeks, and rivers in North America contain multiple species of freshwater snails. Indeed, the southeastern USA has one of the highest freshwater snail diversities in the world. The only habitats typically devoid of snails are ones with high salinity, extremely soft water, or regularly elevated temperatures and habitats that freeze to the bottom. The temperate North American fauna is divided into two main taxonomic groups: (i) caenogastropod snails (formerly called prosobranchs) with about 49 genera and 364 species (Fig. 9a) and (ii) pulmonate snails with 29 genera and 162 species (Fig. 9b). Pulmonate gastropods include more terrestrial species (order Stylommatophora) than aquatic species (order Basommatophora). These groups are easily distinguished by the caenogastropod's possession of an internal gill (ctenidium) and an external operculum which serves as a moderately impenetrable shell door which is attached to the back of the "foot." Once the foot and other soft parts are retracted into the shell, the snail is protected from many aquatic predators and temporarily inhospitable environments. Pulmonate snails, which lack both a gill and an operculum, have secondarily re-invaded aquatic habitats from terrestrial environments and have retained from that terrestrial origin a mantle cavity which functions as a pseudo-lung as long as moist conditions are maintained in the cavity. These snails can climb out of water temporarily to either obtain atmospheric oxygen or avoid aquatic predators.

B. Distribution and Characteristics of Selected Groups

While some large groups of snails are widely distributed in North America, others are confined to limited regions of the USA and Canada.

Among the caenogastropods, members of the family Pleuroceridae are most abundant in rocky headwater streams through large rivers of the southeastern USA and lower Midwestern states. The genus *Juga* is the only pleurocerid found west of the Rocky Mountains. Pleurocerid snails tend to be relatively large snails, and some are heavily armored. Adults are slow at dispersing, which has contributed to severe range reductions in some species that are confronted by dams, pollution, and invasive species. *Goniobasis* is the most diverse genus of pleurocerid snails with over 100 recognized species.

The family Ampullariidae contains the largest of our snails, commonly called apple snails. Native to southern Florida and the American tropics, they are popular in the aquarium industry and many species have been introduced in the USA. Apple snails are a pest of taro crops in Hawai'i. One apple snail species, *Pomacea paludosa*, may get softball sized, and another, *Marisa* sp., has a large ramshorn-shaped shell.

© 2011 Elsevier Inc. All rights reserved.
ISBN 978-0-12-381426-5, DOI: 10.1016/B978-0-12-381426-5.00009-0

FIG. 9A. The Japanese pond snail (*Bellamya* sp.), an example of a caenogastropod, is a common invasive species.

FIG. 9B. The pond snail *Lymnaea* sp., an example of a pulmonate gastropod.

Snails in the family Viviparidae are large bodied and widespread both worldwide and in the eastern states and provinces of North America. The life cycles of these sexually dimorphic snails vary latitudinally from about 2 years in northern areas to 1 year in southern states. Two common genera are *Campeloma* and *Viviparus*, both of which are found in many rivers and lakes. The Asian genus *Bellamya* has been widely introduced in North America.

The family Hydrobiidae, whose members are relatively small and thick-shelled, is the most diverse and widespread mollusc family in freshwaters of North America. In addition to dominating the snail fauna of northern lakes, this family also contains some estuarine and marine species. Identifying these snails by shell alone is nearly impossible, and therefore the structure of the penis is required for species classification. Some common genera are *Amnicola* (only 8 species currently recognized but found in over 40 states and southern Canadian provinces), *Somatogyrus, Marstonia,* and *Pyrgulopsis*. Hydrobiids in the drier western states are often endemic to isolated groundwater seeps and springs. The invasive New Zealand mud snail (*Potamopyrgus antipodarum*) has been widely introduced in the western USA, and threatens native snail populations.

The habitat, diversity, and distribution of pulmonate snails tend to be different from those of caenogastropods because of their shorter life cycles, hermaphroditic reproduction, greater passive dispersal by vertebrates, and potential for air-breathing in otherwise low-oxygen conditions of both lentic and lotic habitats. Three prominent groups are the lymnaeids, physids, and planorbids.

Snails in the family Lymnaeidae are large individuals with broad triangular tentacles. They are common worldwide and represent the greatest diversity of pulmonates in Canada and the northern USA with around 55 species. They range in size from the giant *Lymnaea megasoma* and *Lymnaea stagnalis*, with lengths of about 5 cm, to the relatively small *Lymnaea obrussa*, at about a centimeter in length. Lymnaeids can be found from swampy meadows and ephemeral ponds to permanent ponds and small streams. The common *Lymnaea columella* can be collected just above the water surface in many habitats and is recognizable by its large aperture and almost transparent shell. The lymnaeid genera *Lanx* and *Fisherola* are limpets, with shells shaped like low cones. They can be distinguished from planorbid limpets by their flat triangular tentacles. Planorbid tentacles are long and thin.

Some of the most common and widely collected pulmonates throughout North America are in the family Physidae. They can be recognized by their small, sinistral (see Section IIA) shells with elevated spires. Physids occur in temporary ponds through permanent lakes and in small streams through large rivers (but then usually in slowly moving or static side channels). Their dispersal rates by crawling and transportation by vertebrates probably exceed those of all other snails. Their actual species diversity is controversial at present because past species designations based on shell morphology have been challenged by both recent molecular evidence and natural history observations showing that the shape of the shell can vary considerably in populations of a single species. Molecular systematists have recently shown that only about six valid species exist out of the hundreds that were originally described based on apparent differences in the shells.

The last common and fairly diverse group of pulmonates is composed of the ramshorn snails in the family Planorbidae, including the common genus *Helisoma*. In general, they are easily distinguished by their planospiral shells (see Section II.A), which range in size from 1 to 30 mm in North America. Members of the moderately diverse planorbid freshwater limpets are easily distinguished from other caenogastropods by their simple cone-shaped shells and their narrow, thin tentacles. The stream-lined shape is ideal for attaching to rocks or plants in turbulent streams.

C. Evolution, Invasion, and Extinction

The ancestors of all freshwater snails are marine species of *Caenogastropoda*. One group (freshwater *Caenogastropoda*) evolved continuously in aquatic environments via a marine to estuarine to freshwater pathway, whereas another group (Pulmonata) evolved indirectly through a marine to intertidal (probably) to terrestrial to freshwater pathway. Freshwater caenogastropod species are typically more localized in distribution and thus have had more opportunities to become genetically isolated and radiate into new species. Unfortunately, this isolation also increases the chances of extinction for individual species in this group. Indeed, 45–75% of the species in the two largest families (Hydrobiidae and Pleuroceridae) are considered at risk. By contrast, pulmonates are less diverse in part because their opportunities for genetic isolation are reduced by the passive dispersal of adults or eggs by birds and insects.

The ecological functioning of aquatic and some terrestrial communities are threatened by both loss of native species and invasion of exotic species. Two of the worst offenders are the channeled apple snail, *Pomacea canaliculata* (family Ampullariidae), and the New Zealand mud snail, *Potamopyrgus antipodarum* (family Hydrobiidae), but a total of 37 exotic snail species have invaded North American freshwaters. The channeled apple snail, which is found in Florida and some nearby southeastern states, competes with a native apple snail and is wreaking havoc on native aquatic plants. The mud snail has reached levels of animal production in the Snake River of Idaho that are higher than ever recorded for an aquatic invertebrate. In the process they are reducing primary production by 90% and altering the flow of energy and inorganic nutrients.

II. FORM AND FUNCTION

A. Anatomy and Physiology

FIG. 9C. The freshwater limpet *Lanx* features a conical shell.

The most prominent feature of a snail is its external shell, which not only protects the snail from predators and inhospitable environments but also serves as a handy character to identify species. Shells come in three basic shapes: (i) simple flattened cone (Fig. 9c), as in the freshwater limpet species of *Ancylus*—a group unrelated to marine limpets; (ii) planospiral (whorls of the shell confined to one plane), as in the common pulmonate pond species of *Helisoma* (Fig. 9d); and (iii) elevated spire, as in another widespread pulmonate *Physa* (Fig. 9d) as well as in most freshwater caenogastropods, like *Goniobasis* and *Lymnaea* (Fig. 9b). Spiral shells are supported by a central column

FIG. 9D. The ramshorn snail *Helisoma* bears a planospiral shell.

which also provides an attachment site for muscles supporting soft body parts.

Shells are composed of an outer mostly protein-based periostracum overlying a thick, major support layer of mostly crystalline calcium carbonate. The entire shell is secreted by the fleshy outer body layer of the snail using organic material and inorganic $CaCO_3$ ions absorbed from the water and food. In very soft waters, $CaCO_3$ may be so rare that the mollusc population will be sparse or absent. Many pulmonates have thin shells, while the shells of caenogastropods tend to be thicker, especially in species living in large rivers with big predators. Crayfish can chip away the fragile walls of a pulmonate snail, but it takes progressively bigger predators to crush the shells of caenogastropods, especially those of thick-walled river snails.

Shell shape is important in identifying snails, especially at higher taxonomic levels. Some important terms can be understood by laying the shell of a species with an elevated spire so that the opening (aperture) is up and the shell spire and its apex (pointed end) extend away from you. If the aperture now opens toward the left, the shell is termed sinistral, as occurs in the physid snails. If instead the opening is on the right, you are viewing a dextral shell which characterizes many groups, including the lymnaeids and all North American caenogastropods. Many other characters are used to describe shells, including the length of the shell, shape of the body whorls and channel depth between them, shape of the lip of the aperture, types and numbers of spines and ridges, and growth lines on the operculum. In the last case, these lines are termed concentric if successive ones entirely enclose the previous lines, or they are called multi-spiral or pauci-spiral if arranged in a spiral.

The shell can enclose the entire soft body parts of the snails, which consist of the (i) head and two or four tentacles (with eyes at the base of the primary pair); (ii) muscular foot; (iii) visceral mass (mostly digestive, excretory, and reproductive organs); and (iv) overlying mantle, which secretes the shell and in caenogastropods forms an anterior cavity enclosing the gill. The eyes detect light intensity but can see only poorly-resolvable images. Olfactory detectors are well developed because these provide the primary information on food sources.

The primary functions of the foot are movement, sensory reception, and food acquisition. Snails move using a combination of cilia and muscles. The cilia propel the snail through a mucous layer laid down by glands, while the foot muscles push the snail forward with contracting muscle waves. Snails acquire food using rasping movements of the file-like radula, which is in the snail's mouth (somewhat similar to a tongue). The radula can be projected forward or withdrawn into the mouth. Food particles are scraped from the bottom and enter the snail as the radula is withdrawn.

B. Reproduction and Life History

Snails are notable for their diversity of life history patterns. Pulmonate snails tend to reproduce once (semeloparity) and live no more than a year, whereas caenogastropods commonly reproduce multiple times (iteroparity) and have a life cycle of up to 4–5 years. Semelparous species tend to devote a greater portion of their total energy accumulated to reproduction than do iteroparous taxa. However, the frequency and amount of reproduction seem related to food availability and environmental conditions like water hardness and ambient temperatures.

Depending on the species, snails may have separate sexes (dioecious), hermaphroditic (the same individual is a male and a female sequentially or simultaneously), or parthenogenetic (males are absent or very rare). Hermaphroditism and parthenogenesis in snails are life history strategies helpful in overcoming problems of finding mates when the snail finds itself alone or at very low densities, such as when isolated in a newly colonized pond or a small headwater stream. Eggs are laid externally on a suitable substrate or within the mantle fold where the young hatch.

Caenogastropods are typically dioecious, though members of the family Valvatidae are hermaphroditic and some families have parthenogenetic members, including the invasive New Zealand mud snail (Hydrobiidae) and the common river snail *Campeloma* (Viviparidae). Most males use the enlarged right tentacle as a copulatory organ (in Viviparidae) or possess a specialized copulatory penis.

Pulmonates, in contrast, are hermaphroditic. The ova are fertilized either by the same individual's sperm or as a result of copulation with another snail. Egg production is higher, however, when cross fertilization occurs. When a pulmonate has stored a large amount of sperm, it adopts the role of the male and seeks a mate with temporary female proclivities.

III. ECOLOGY, BEHAVIOR, AND ENVIRONMENTAL BIOLOGY

A. Trophic Position

Freshwater snails are primarily herbivores or detritivores, but some species incidentally consume microinvertebrates while scraping algae from benthic surfaces, and other species are known to scavenge occasionally on carrion. Snails in the family Viviparidae are suspension feeders. A diet that includes some animal material is known to improve growth rates in lymnaeid snails. However, unlike some of their marine relatives and a few tropical freshwater species, no North American freshwater snails are considered predaceous. An algal diet is generally more nutritious than a diet of dead organic matter (detritus) unless the microbial content on the detritus is high. Nonetheless, both physid and planorbid pulmonates often seem to prefer detritus over algae as a regular food source. All these food sources are detected primarily by chemoreception.

Benthic algae are a primary food source for most snails because these autotrophic organisms are frequently abundant, generally nutritious (low carbon to nitrogen ratio), and never run away or fight back! Cyanobacteria (bluegreen algae) and various forms of long filamentous green algae are avoided by many species because they may lack vital nutrients, contain toxins, or are merely too large to eat. The size and type of algae consumed are partially determined by the form of the radula. For example, the radular teeth of lymnaeids are relatively simple but large, making them effective for grazing on long strands of filamentous algae, whereas the small teeth of physids are better for scraping tightly attached diatoms from rocks. An unusual feeding strategy occurs in the caenogastropod *Bithynia tentaculata*, which not only grazes on benthic algae but also employs its gill to obtain algae suspended in the water. If algae are not sufficiently abundant, snails will eat aquatic weeds, which are less nutritious and sometimes toxic.

B. Biotic Interactions and Population Limiting Factors

The role of predators versus competitors (mostly for food) in controlling density and diversity of snail population remains controversial and probably varies among habitats and over time. Below is an introduction to this subject.

Competition for Food: Various field experiments have demonstrated that food abundance affects snail survival and fecundity and that snails can also alter algal densities through intense grazing, thereby increasing potential competition among snails. Competition for food may be within or between snail species or between snails and another type of invertebrate (e.g., grazing mayflies or caddisflies). Studies of intense grazing under natural conditions as well as field experiments have demonstrated that snails can alter the biomass, production, nutrient content, species diversity, and favored growth form of algae, especially in low-shade conditions where light does not limit algal growth. By contrast, low levels of grazing can actually stimulate algal production. The species of snail and the intensity of grazing influence the relative abundance of algae.

Predation on Snails: Vertebrate and invertebrate predators seem to strongly affect population densities and community diversity of gastropods, as evidenced by species distributions, field experiments, and the presence of apparent anti-predator adaptations in snails. The last includes (i) thick shells and armored spikes or ridges to escape shell-crushing predators; (ii) behaviors like shell-shaking to throw off or discourage a predator; and (iii) crawling out of the water to escape invertebrate predators or seeking underwater refuges.

Snail-eating fish (molluscivores) are common in rivers and lakes. These fish either break the shell or swallow the snail whole. Many species of sunfish are prominent molluscivores including the redear or shell-cracker sunfish (*Lepomis microlopus*) and the pumpkinseed sunfish (*Lepomis gibbosus*), which prefer large, thin-shelled snail species. Other fish, such as the central mud minnow (*Umbra limi*), prey on snail eggs and juveniles in permanent ponds and can limit the distribution of some lymnaeid snail species. Predation from fish in lakes limits most thin-shelled snail species to refuges provided by stands of aquatic weeds, while thicker-shelled species like the planorbid *Helisoma* and various caenogastropods can survive in open sandy and rocky areas. Fish predation can be a selective force influencing shell shape and shell thickness. Fat shells are more effective against fish, while relatively long shells reduce predation from shell-invading predators like crayfish. The shells of physid snails tend to be thinner in very small streams, wetlands, and ephemeral ponds, where fish predation is minimal or absent. Molluscivores can alter feeding behaviors of snails and may cause them to stay within refuges until starvation forces them out (fish predators) or to avoid foraging areas in general where the predator is known to frequent (e.g., cobble areas where crayfish hide). The ability to easily find shelter seems as important a survival strategy as having thicker shells. The presence of these predators is associated with increased abundance of algae because of the absence of large numbers of foraging herbivorous snails.

Invertebrate predators on adult snails either break the shell (crayfish) or invade the shell to gain access to soft body parts. These include crayfish, predaceous diving beetles (e.g., *Cybister*), giant water bugs (Belostomatidae), and carnivorous leeches (e.g., *Nephelopsis obscura*). If shell-breaking is an option, thin-shelled species are favored by the predator, whereas shell-invading predators are not affected by shell thickness but may be hindered by an aperture blocked by an operculum. Crayfish are very efficient predators when not themselves threatened by fish and are known to eat 100 or more snails a night. This intense predation can shift the species composition of lakes toward larger and/or thicker-walled species of snails.

Parasitic Control: Scientists know a relatively large amount about the role of parasites in controlling snail populations because many of the same parasites also infect humans at a different life stage. Snails are often infested with trematodes (parasitic members of the flatworm phylum Platyhelminthes) beginning with larval stages. Trematode infections in snails can accelerate or depress growth rates, alter and even halt egg production through consumption of the snail's reproductive organs, and change foraging times of the snail (affecting chance of consumption by the next host). The degree of trematode infestation varies among ponds and years, with infection rates of *Lymnaea elodes* ranged in one study from insignificant to nearly half the population.

C. Environmental Physiology

Snails in general and pulmonates in particular encounter a wide variety of environmental conditions which may promote growth or depress populations. In the temperate zone, most pulmonate snails can tolerate temperatures hovering near freezing (0°C) for several months, allowing them to get a jump on breeding in early spring, but they tend to be more common in waters with moderate temperatures. They generally grow faster at higher temperatures, but mortality rises above 30°C. In accordance with their typical habitats, pulmonates are more tolerant of temperature swings than are caenogastropods. If temperatures become excessive, pulmonates will secrete a mucous covering over the aperture to retard water loss. Caenogastropods either tolerate low oxygen conditions or die, while pulmonates have the option of moving out of the water for short periods to obtain atmospheric oxygen. All snails must absorb calcium carbonate to build their shells, and this can limit their distribution. Nearly half the species of snails in North America are limited to medium-hard water (>25 mg/L), while almost all require water at least 3 mg/L (very soft water) or higher. In soft-water habitats, snail shells tend to be thinner, thus reducing protection from predators.

IV. COLLECTION AND CULTURING

Snails can be collected either by sweeping a fine mesh net (1-mm mesh) through aquatic vegetation or by examining rocks, aquatic weeds, and wood snags by hand. Most snails are easy to see with the naked eye, but a few groups are small enough that use of a hand lens is required. If the sample contains a large amount of sand or silt, it may be necessary to sieve the samples first before extracting the snails. As for most invertebrates, sorting samples in a white tray is a helpful way to spot the target species.

Many snail species can be easily cultured in a home or school aquarium as long as the water is well oxygenated, kept at room temperatures (or even slightly cooler at 15–20°C), appropriate food is available (algae or microbially conditioned, organic detritus), and snail-eating fish (like many sunfish) are absent. The very common, herbivorous species can eat algae growing on the aquarium surfaces or on algal-covered rocks added from a stream or pond. Species feeding on organic detritus can be fed leaf litter that has been colonizing in the oxygenated waters of a stream or pond for at least 2 weeks. Avoid fouling the water with too much decomposing material. The snails will continue to grow and reproduce as long as population densities are not too high.

V. REPRESENTATIVE TAXA OF SNAILS AND LIMPETS: CLASS GASTROPODA

A. Family Neritidae

Nerite or Marble Snail (Fig. 9e): Neritina sp.

Identification: Shell thick, heavy, and almost spherical. Shell may be black with small white spots or can be brown, gray, or tan with dense transverse, thin, black lines. Animal black, gray, brown, or mottled, with one pair of slender tentacles. Operculum present. **Size:** 1–10 mm. **Range:** Coosa River, Alabama, and the Gulf Coast to Florida. **Habitat:** Occurs in brackish water, estuaries, and tidally influenced freshwater. **Remarks:** One species, *Neritina reclivata*, is common in the aquarium trade.

B. Family Viviparidae

Viviparus sp. (Fig. 9f)

Identification: Shell globose, brown to black, or some species white to tan with stripes. Animal brown or mottled, with one pair of short, thick tentacles. Operculum present and circular. **Size:** 1–30 mm. **Range:** Midwestern and southern USA. **Habitat:** Ponds, lakes, bogs, and slower portions of rivers and streams, in soft sediments. **Remarks:** These animals are detritivores, although some species filter feed. Produce live young.

Campeloma sp. (Fig. 9g)

Identification: Shell brown to yellow or green, globose, never with stripes. Animal brown or mottled, with one pair of short, thick tentacles. Operculum present and oval. **Size:** 1–25 mm. **Range:** Eastern portions of Canada and the USA. **Habitat:** Ponds, lakes, bogs, and slower portions of rivers and streams, in soft sediments. **Remarks:** These animals are detritivores. Produce live young. Like all viviparids they can live for as long as 5 years.

Livebearing Pond Snail (Fig. 9h): Bellamya sp.

Identification: Shell brown to black, globose, never with stripes. Animal brown or reddish, with two pairs of short, thick tentacles. Operculum present and oval. **Size:** 1–45 mm. **Range:** Native to China and Japan. Widely introduced in North America. **Habitat:** Ponds, lakes, bogs, backwaters, and sloughs, in soft sediments. **Remarks:** These animals are filter feeders, and produce live young. This species was introduced to North America through the pet industry, and can be found in most temperate regions.

C. Family Ampulariidae

Apple Snails (Fig. 9i): Pomacea sp.

Identification: Shell brown to yellow, sometimes with stripes. Animal black, white, yellow, brown, gray, or mottled, with two pairs of short, thick tentacles. Operculum present and oval. **Size:** 1–100 mm. **Range:** Florida, Georgia, Alabama, and Mississippi, USA. Two native species, several tropical species have been introduced. **Habitat:** Ponds, lakes, bogs, rivers, and streams. Eggs are deposited above the water surface. **Remarks:** These animals are omnivores, consuming mostly plant material. These snails are the primary food item for the snail kite. Common aquarium pets.

Giant Ramshorn Snails (Fig. 9j): *Marisa sp.*

Identification: Shell discoidal, colored brown to white, yellow, or green, often with stripes. Animal brown or mottled, with one pair of short, thick tentacles. Operculum present and oval. **Size:** 1–50 mm. **Range:** Native to tropical America, widely introduced in the USA. **Habitat:** Ponds, lakes, bogs, rivers and streams, and hot springs. **Remarks:** Predatory on other snails and will consume plant materials. Common aquarium pets.

D. Family Thiaridae

Cornucopia or Melania Snails (Fig. 9k): *Melanoides sp.*

Identification: Shell elongate cornucopia shaped, colored yellow to brown, rarely black, often with ridged sculpturing, and with red or brown markings. Aperture and operculum oval in shape. Animal brown or black, with one pair of slender tentacles. **Size:** 1–40 mm. **Range:** Introduced throughout the USA, established in the south and in western hot springs, as well as in California, south from the Sacramento River. **Habitat:** Rivers, streams, and hot springs. **Remarks:** This is a popular aquarium snail that is parthenogenic. Several species are common in pet shops.

Tarebia granifera (Fig. 9l)

Identification: Shell cornucopia shaped, brown in color, with nodulose sculpturing. Aperture and operculum oval in shape. Animal brown or black, with one pair of slender tentacles. **Size:** 1–30 mm. **Range:** Introduced in the southeastern USA and in western hot springs. **Habitat:** Rivers, streams, and hot springs. **Remarks:** This is a common aquarium snail.

E. Family Pleuroceridae

Juga sp. (Fig. 9m)

Identification: Shell cornucopia shaped, colored brown to black, often with sculpturing in the form of ridges. Aperture oval in shape, without a channel. Animal brown, with one pair of slender tentacles. Operculum present and oval. **Size:** 10–25 mm. **Range:** Pacific northwestern USA. **Habitat:** Rivers, streams, and hot springs. **Remarks:** Probably several undescribed species. This is the only pleurocerid genus in the West.

Goniobasis sp. (Fig. 9n)

Identification: Shell cornucopia shaped, colored brown to black, often with white or cream stripes. Aperture oval, without a canal. Animal brown, with one pair of slender tentacles. Operculum present and oval. **Size:** 15–30 mm. **Range:** North America, east of the Rocky Mountains in southern Canada and the USA. **Habitat:** Rivers and streams. **Remarks:** Tend to be in faster flowing streams near mid channel.

Goniobasis sp. (Fig. 9o)

Additional Remarks: Demonstrates the range of variation within this genus. This species has broader stripes and lacks the sculpture on the spire.

Pleurocera sp. (Fig. 9p)

Identification: Shell cornucopia shaped, colored brown to black, often with white or cream stripes. Aperture rectangular in shape, with a canal or slit. Animal brown, with one pair of slender tentacles. Operculum present and oval. **Size:** 20–40 mm. **Range:** North America, east of the Rocky Mountains in southern Canada and the USA. **Habitat:** Rivers and streams. **Remarks:** Occurs in slow flowing areas near watercourse margins and under banks.

Leptoxis sp. (Fig. 9q) and Athearnia sp.

Identification: Shell elongated, top shaped, or subglobular, colored brown to black, sometimes with faint stripes. Aperture round or oval, without a canal or slit. Animal brown, with one pair of slender tentacles. Operculum present and oval. **Size:** 10–25 mm. **Range:** East of the Rocky Mountains in the USA. **Habitat:** Rivers and streams in fast flowing water. **Remarks:** One species is federally listed as endangered. The genus *Athearnia* is probably a synonym.

Spiny River snails: Io fluvialis (Fig. 9r) and Lithasia sp.

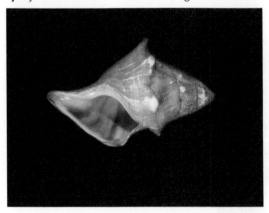

Identification: Shell elongated, with large lateral knobs or projections, colored brown to tan. Aperture angular, with or without a long canal projecting anteriorly. Animal yellow, with one pair of slender tentacles. Operculum present and oval. **Size:** 10–50 mm. **Range:** Ohio, Tennessee, Black, Spring, and Big Black river systems (*Lithasia*). Tennessee River system, extirpated from most of its range (*Io*). **Habitat:** Large rivers, in fast flowing, well-oxygenated portions. **Remarks:** This species is protected as threatened.

F. Family Bithyniidae

Bithynia sp. (Fig. 9s)

Identification: Shell small, globose, colored brown to black. Aperture oval, with a white operculum. Animal to brown or black often with gold speckles, bearing one pair of tentacles. **Size:** 1–14 mm. **Range:** Native to Eurasia, introduced widely in the Great Lakes and mid-Atlantic regions. **Habitat:** Lakes, ponds, and slow streams and rivers, particularly in systems high in calcium. *Remarks: Bithynia* can feed both on algae scraped from the substrate and by filtering material from the water column.

G. Superfamily Hydrobioidea

Hydrobioid Snails (Fig. 9t; the following taxa shown clockwise from top): Flumini-cola sp., Amnicola sp., Pristinicola sp., Notogillia sp., Tryonia sp., Pyrgophorus sp., and Littoridinops sp.

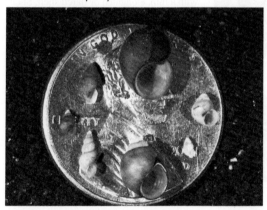

Identification: Shell small, usually elongated, conical or globular, may rarely be discoidal shaped, colored brown to black. Aperture oval to round. Animal yellow to brown or black, with one pair of slender tentacles. Operculum present and oval. **Size:** 1–4 mm, rarely as large as 9 mm. **Range:** Throughout North America, outside the Arctic. **Habitat:** Seeps, springs, streams, ponds, lakes, and rivers. **Remarks:** nearly 50 genera with around 500 described and undescribed species. Some species are protected by state or federal law.

New Zealand Mud Snail (Fig. 9u): Potamopyrgus antipodarum

Identification: Shell elongated, colored gray, brown to black. Rarely with small spines. Aperture oval. Animal gray or black, with one pair of slender tentacles. Operculum present and oval. **Size:** 1–4 mm. **Range:** Endemic to New Zealand, introduced in many places in temperate North America. **Habitat:** Small to large streams and rivers, sometimes estuaries. **Remarks:** Nonnative invasive species that has displaced some native taxa. Reproduces parthenogenically (all clones of the mother). Males rare.

H. Family Valvatidae

Valvata sp. (Fig. 9v)

Identification: Small snails with a depressed shell, often with one, two, or rarely three ridges. Shell color yellow, gray, brown, or black. Aperture circular, with a circular operculum. Animal gray or black, with one pair of slender tentacles, and a single, feather-like gill carried outside the shell when the animal moves. **Size:** 1–8 mm. **Range:** Temperate North America. **Habitat:** Small to large streams, rivers, and ponds sometimes in lakes, usually in aquatic plants. Prefers water with a neutral pH. **Remarks:** One species is protected under federal law.

I. Family Lymnaeidae

Pond Snails (Fig. 9w): *Lymnaea sp.*

Identification: Small to very large snails with a thin, elongated or flared shell. Shell color yellow or gray to brown. The shell always spirals to the left. Aperture usually oval, or circular, never with an operculum. Animal gray or yellow, with one pair of triangular tentacles. **Size:** 4–50 mm. **Range:** Temperate North America. **Habitat:** Slow streams and rivers, very common in ponds and lakes, often on muddy bottoms, sometimes in aquatic plants. **Remarks:** Some authors have split this genus into several genera; however, it is probably unwarranted. One species is protected under federal law.

Lymnaea sp. Fig. 9x

Additional Remarks: Eurasian invasive species, widespread.

Lymnaea sp. Fig. 9y

Additional   Remarks: Ubiquitous wetland species.

Lymnaea sp. Fig. 9z

Additional   Remarks: Extremely elongated species.

Lymnaea sp. Fig. 9aa

Additional Remarks: Found in large northern lakes.

Western Limpets: *Lanx sp. (Fig. 9ab) and Fisherola sp.*

Identification: Shell cap shaped, never spiral. Shell color gray to brown, often with red or yellow mottles. Operculum always absent. Animal gray or brown, with one pair of triangular tentacles. **Size:** 4–18 mm. **Range:** Western North America. **Habitat:** Streams and rivers, often in riffles attached to rocks. **Remarks:** The cap shape of the shell allows the animal to occupy swift waters, giving the current little purchase on the shell.

J. Family Physidae

Tadpole or Bladder Snails: *Physa sp. (Fig. 9ac) and Aplexa sp.*

Identification: Small snails with a thin, matte or shiny, cylindrical shell. Shell color is gray to brown. The shell always spirals to the right. Aperture is oval, never with an operculum. Animal gray, yellow, or black with one pair of slender tentacles. **Size:** 1–25 mm. **Range:** Temperate, subarctic, and subtropical North America. *Physa* is cosmopolitan. *Aplexa* is limited to northern USA and Canada. **Habitat:** Very common in ponds, lakes, slow streams, and rivers. Occasionally in swift water. **Remarks:** This genus has been divided into numerous genera and species; however, molecular analyses demonstrate that there are actually very few variable species all in one genus.

K. Family Planorbidae

Ramshorn Snails: *Helisoma sp. (Fig. 9ad), Drepanotrema sp., and Biompholaria sp.*

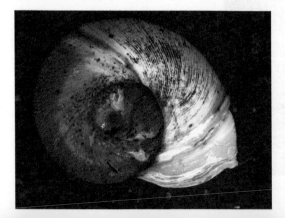

Identification: Small- to medium-sized snails with a discoidal shell. Shell color yellow or brown. Aperture subcircular, typically in line with the plane of the shell, never with an operculum. Animal gray, brown, or black, with one pair of slender tentacles. **Size:** 1–30 mm. **Range:** Widespread in North America. **Habitat:** Ponds, lakes, and slow portions of streams and rivers, usually in aquatic plants. Feed on bacteria. **Remarks:** Some tropical species carry fluke worms. *Helisoma* includes species previously contained in *Planorbella* and *Planorbula*.

Ramshorn Snails: Helisoma sp. (Fig. 9ae), Drepanotrema sp., and Biompholaria sp.

Additional Remarks: Example of variation in shell morphology.

Carinifex newberryi (Fig. 9af)

Identification: Medium-sized to very large snails with a shoulder ridge or keel, more pronounced in animals from alkaline habitats. Shell color yellow to brown. Aperture subcircular, not in line with the plane of the shell, never with an operculum. Animal gray or brown, with one pair of slender tentacles. **Size:** 4–50 mm. **Range:** Great Basin Desert and southern Cascades Mountains. **Habitat:** Large lakes, prefers semi-alkaline lakes with springs. **Remarks:** Included in the genus *Helisoma* by some authors. This taxon may include several species.

Gyraulus sp. (Fig. 9ag), Menetus sp., Promenetus sp., and Micromenetus sp.

Identification: Small snails with a discoidal shell. Shell color white, yellow, or brown. Aperture subcircular, typically out of line with the plane of the shell, never with an operculum. Animal gray, brown, or black, with one pair of slender tentacles. **Size:** 1–8 mm. **Range:** Widespread in North America. **Habitat:** Quiet waters in ponds, lakes, and backwater portions of streams and rivers. **Remarks:** Some species occur in temporary ponds.

Gyraulus sp., Menetus sp., Promenetus sp. (Fig. 9ah), and Micromenetus sp.

Additional Remarks: Example of variation in shell morphology.

Vorticifex sp. (Fig. 9ai)

Identification: Small snails with a highly flattened, discoidal shell that rapidly expands in width as it spirals to the aperture. Shell color yellow or brown. Aperture subcircular, never with an operculum. Animal gray, brown, or black, with one pair of slender tentacles. **Size:** 8–30 mm. **Range:** California, Oregon, and Nevada. **Habitat:** Streams and rivers. **Remarks:** Common in well-oxygenated riffles.

Planorbid and Acroloxid Limpets: Ferrissia sp. (Fig. 9aj), Rhodacmea sp., Laevipex sp., Hebetancylus sp., and Acroloxus sp.

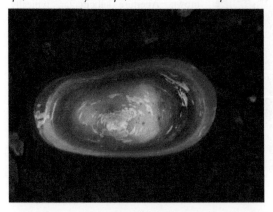

Identification: Shell cap shaped, never spiral. Shell color gray, yellow, or brown. Operculum always absent. Animal gray or brown, with one pair of thick tentacles. **Size:** 1–8 mm. **Range:** Widespread in North America. **Habitat:** Streams and rivers, often in riffles attached to rocks.
Remarks: These limpets are more elongate than their Lymnaeid relatives.

Chapter 10

Mussels and Clams: Phylum Mollusca, Class Bivalvia

I. INTRODUCTION, DIVERSITY, AND DISTRIBUTION

A. General Features

In addition to the snails (class Gastropoda) discussed in the last chapter, freshwater molluscs (or mollusks) include a large group of species whose bodies are enclosed within a pair of hinged shells (or valves). These members of the class Bivalvia (sometimes called Pelecypoda) are represented in North America by freshwater pearly mussels or naiads (superfamily Unionoidea), one native species of true mussel in the family Dreissenidae (*Mytilopsis leucophaeata*), and the pill, fingernail, and pea clams (family Sphaeriidae). In addition to these native bivalves, North American waters have been invaded by the Asian clam (Corbiculidae, *Corbicula* spp.), zebra and quagga mussels (Dreissenidae, *Dreissena* spp.), and a number of exotic sphaeriids. North America contains the greatest diversity of pearly mussels in the world. Two families of Unionoidea occur in North America: Unionidae with about 300 species (Fig. 10a) and Margaritiferidae with about 5 species. The family Sphaeriidae (formerly also Pisiidae) contains 4–5 genera with about 40 species.

As in snails, the protective shells of bivalve molluscs are composed of a thick calcareous matrix overlain with a thin, protein-based periostracum. When the valves are partially opened, they allow the animal to acquire oxygen and filter water for food through one or more siphons as well as extend their foot to burrow or obtain benthic food.

All native bivalves in North American freshwaters are adapted to a sedentary life in areas with sufficient water movement to ensure a continued supply of oxygen and food. As a result, most species live in headwater streams through large rivers, although some have colonized large lakes, small ponds, and even ephemeral wetlands (mostly small clams). Unlike ocean clams and mussels which have planktonic larvae that swim in open water, freshwater species either brood their young (clams) or have larvae called glochidia (mussels), which temporarily parasitize fish and at least one amphibian species.

Pearly mussels have a long-term relationship with humans. They have served as (i) food for Native Americans for at least 5000 years (though nowhere near as tasty as marine mussels); (ii) a source of pearly buttons, which were stamped from shells from the late 1800s through early 1900s until the advent and widespread use of plastics in the 1940–1950s; and (iii) modern commercial sources of both freshwater pearls and seed pearls for the cultured pearl marked in Asia.

B. Macro- Through Micro-Scale Distributions

The continental-scale distribution of bivalves in North America reflects in great part the reproductive and dispersal abilities of the bivalve, the barrier of the continental divide, the presence of the

ISBN 978-0-12-381426-5, DOI: 10.1016/B978-0-12-381426-5.00010-7

FIG. 10A. Examples of three freshwater mussels.

Appalachian Mountains, and the extent of Pleistocene glaciation. The last two have been circumvented by many species over time, but the continental divide separating Pacific and Atlantic/Gulf drainages remains a very effective barrier. For example, only 9 species of mussels occur west of the Rocky Mountains, while 295 are found east of the Rockies, with only 2 species shared by both zoogeographic areas. Tennessee, Kentucky, and other southeastern states are the sites of maximum richness and abundance of mussels.

At the watershed scale, longitudinal patterns of bivalve distribution and abundance are closely linked to reproductive strategies and dispersal mechanisms. For mussels, distributional patterns reflect the distribution of host–fish relationships in the connecting network of rivers and streams. Clams, by contrast, tend to be more widely distributed because they do not depend on fish to disperse their larvae, and even adults can be transported externally by birds from one ecosystem to the next. The nonnative zebra and quagga mussels (*Dreissena polymorpha* and *Dreissena bugensis*, respectively) have planktonic larvae that are easily transported downstream. Upstream dispersal is much more problematic and essentially depends in North America on transport on or within barges and pleasure boats. These distribution patterns can also be influenced by large-scale variables like catchment geology (affecting stream bed type, sediment size and transport, pH, and other factors), temperatures, stream size, etc.

The importance of habitat characteristics on distribution, abundance, and diversity of bivalves is controversial. At one time scientists believed that you could predict bivalve habitat by merely knowing some key habitat features, such as stream size, flood patterns, and sediment size. However, even in pristine systems, the story is apparently much more complicated and mussel distributions are difficult to predict. And, of course, human impacts have modified natural patterns greatly, especially by eliminating fish species or blocking the migration of larva-carrying fish. Humans have also had major negative impacts by altering substrate size, sediment turbidity, and general water quality. In general, however, river systems that have relatively stable flow patterns (less flashy) and more diverse habitats will support more mussel species. Likewise, sphaeriid clams do better in permanent ponds, lakes, and river systems, though some species (e.g., *Pisidium casertanum*) occur most frequently in ephemeral habitats. *Corbicula* frequents a variety of soft sediment habitats and tends to live in somewhat deeper waters compared to freshwater mussels. Dreissenid mussels attach to hard substrates and generally avoid silty and sandy substrates unless they find an occasional rock, living or dead mussel, or other hard natural or artificial substrate to colonize.

C. Extinctions, Commercial Harvests, and Invasions

For a variety of reasons—almost all of which are related to human activities—the plight of native bivalves (especially mussels) is arguably worse than any other large group of animals in North America. A 1993 summary of the conservation status of freshwater mussels on this continent identified over 70% of mussel taxa as being endangered, threatened, or species-of-special-concern. In part because of political considerations, this list does not agree with the official list maintained by the US Fish and Wildlife Service—which barely added any species to their lists during most of the first decade of the twenty-first century. However, it is clear that mussel species continue to go extinct no matter what the size of the official list. In 1993, 21 species were recognized as having gone extinct in historical times, and that number has now risen to at least 31 species and 3 subspecies. Although no sphaeriids are currently on the endangered species list, this could reflect the lack of study of this family.

Although many factors have led to extinction of mussel species, two primary causes have been disruption of habitats (e.g., by increased sediment load) and severance of dispersal pathways for glochidia-bearing fish following dam construction. A frightening scenario for the future survival of mussels is that humans, in the name of supposed "green technology," are discussing construction of new dams after several decades of blockage by various federal environmental protection laws!

Some mussel species are also threatened by the search for freshwater pearls or use of shell fragments in commercial production of pearls. Harvesting of freshwater pearls for jewelry was common in the early 1900s and only dramatically slowed in the 1950s when the Japanese developed techniques for culturing the much larger, regular-shaped, and highly prized marine pearls. This industry relies greatly on shell fragments from certain freshwater mussels (particularly *Megalonaias nervosa* and *Amblema plicata*) to serve as seed pearls (nuclei) for insertion between the shell and mantle of an oyster, which leads the marine mollusc to produce a valuable cultured pearl. Freshwater pearls are also produced commercially in smaller quantities in North America and Asia. Many responsible US states are now limiting the kind and amount of mussels that can be harvested or completely banning this practice because of threats to our native mussel fauna.

Another significant or potentially important threat to native bivalves comes from invasive species from other continents. At the current time, North American freshwaters are home to nine exotic species consisting of the now infamous zebra and quagga mussels along with two Asian clams (the widely distributed *Corbicula leana* and *Corbicula fluminea* of the southwestern USA) and at least five sphaeriid species. Scientists lack sufficient knowledge of the normal distribution of native pill, fingernail, and pea clams to determine the effects of invasive species. As far as we know, *Corbicula* does not seem to have been the cause of any extirpations of native species, but it could have influenced their densities from competition for algal food. By contrast, zebra mussels have had very demonstrable negative effects on native mussels and to a lesser extent on snails. The negative impacts are associated with overgrowth on shells of the native molluscs and general food competition (with mussels). Both *Corbicula* and *Dreissena* have had marked negative effects on human industry and public works, especially by clogging water intake pipes.

II. FORM AND FUNCTION

A. Anatomy and Physiology

The anatomy of bivalve molluscs resembles that in gastropod molluscs except that the body has not undergone torsion and spiraling (eliminating one half of paired organs) and the shell is hinged rather than single and spiraled. Like snails, bivalve use a single foot in locomotion, but unlike gastropods, bivalves lack a head. This lack of cephalization is associated with a sedentary life-style.

As in snails, the shell is secreted by the underlying mantle tissue and is formed of crystalline calcium carbonate overlain by periostracum that helps retard dissolution of the shell, especially in soft water. The inner shell has articulating "teeth" and an elastic dorsal ligament, which in combination with prominent dorsal anterior and posterior adductor muscles allow the shell to spring open or be closed tightly. The inner surface of the muscle shell is formed by a pearly nacreous layer which is often white but may be colored in various shades of copper, pink, blue, or violet, etc., depending upon the species. This nacre has a long association with humans because of its economic value. It formed the pearly coating of buttons stamped out by humans until about 1950. Furthermore, when a grain of sand or other particle (including inserted bits of mantle taken from a conspecific mussel) is positioned between the shell and the underlying mantle, the latter coats the irritating object with nacre, producing a freshwater pearl or a marine cultured pearl. The shape of the shell, external features, hinge shape, and nacre color are used to various degrees in bivalve classification.

The paired shells enclose the enfolding mantle and visceral mass. The latter includes labial palps (useful in sorting edible food and inedible particles as pseudofeces), paired ctenidia, and other internal organs. The ctenidia serve as both respiratory gills and sites of ion exchange, suspension feeding, and larval brooding.

Native bivalve molluscs are sedentary but can burrow vertically and move along the substrate short distances using the combined muscular and hydrostatic action of the fluid-filled, hatchet-shaped foot. When extended vertically, the foot also serves as an anchor for the mussel. This locomotion contrasts with the truly sessile dreissenid mussels and some juvenile unionid mussels which attach to the substrate with tough fibers called byssal threads. Zebra and quagga mussels can cut the byssal threads and move short distances in some cases, but this is not equivalent to the more frequent locomotion of the native bivalves. The strength of these fibers enables clumps of dreissenids to clog pipes and be very difficult to remove.

Balancing water influx and loss of salts is a critical task for mussels because of the dilute medium in which they live. These conditions pose special problems for molluscs because of the need to extract calcium carbonate for shell construction. If faced with extreme conditions, such as exposure to air, mussels close the shell, slow their metabolic rate, and extract ions from the shell as needed. They can stay closed for hours to days depending on the species and ambient temperatures. In general, the average salt concentration in the internal fluids of mussels is up to one-half that found in other freshwater invertebrates.

B. Reproduction and Life History

Life cycles among families of freshwater molluscs differ considerably in length and complexity. Unionid mussels live for as short as a few years to as long as 200 years in rare cases, with several decades probably nearer the average. By contrast, the maximum life span of some sphaeriid species may be as short as several months in ephemeral habitats and as long as a couple of years in permanent lakes. Asian clams survive up to about 7 years, while dreissenid mussels live for a maximum of about 2–3 years.

The gonads of bivalves are paired, large, and often associated with the gut and digestive gland. The testes are typically whitish, whereas ovaries are pinkish-brown. Males broadcast spermatozoa (packaged as "sperm balls" in mussels), but fertilization occurs in the mantle cavity in all North American bivalves except the exotic dreissenids in which the ova are released for synchronized external fertilization. Most mussels are iteroparous over their lives but spawn only once per year. The age of first reproduction varies considerably with the life span of the adult, with some species maturing within a year and others delaying as long as 20 years.

Freshwater bivalves evolved from marine ancestors with the general characteristics of separate sexes and planktonic larval stages, including a shelled veliger stage. From that beginning, freshwater bivalves in North America have dispensed with a veliger larvae (except in the exotic dreissenids) and many have become hermaphroditic, especially sphaeriids and Corbicula, in which it is the predominant form. Only around 2% of pearly mussels are exclusively hermaphroditic, though perhaps 10% of the species are occasionally dual-sexed. Male–female ratios are often highly skewed from a 1:1 sex ratio, possibly because hormones may play a role in sex determination. Rather than having planktonic larvae like their marine ancestors, mussels produce, brood, and then release large numbers of ectoparasitic larvae (glochidia) in each breeding season; these survive weeks to months on their host before metamorphosing into the adult form. Glochidia are produced in several shapes (one type per species) depending on whether they are meant to attach to the gills or the external body and fins of fish (or to a salamander host in one case). Sphaeriids and Asian clams, by contrast, brood their offspring to a relatively large shelled juvenile stage, which resemble miniature adults.

To increase chances of finding the correct host, a mussel species can broadcast large number of glochidia, or package and release the fertilized eggs and larvae as conglutinates that resemble a food item for fish, or even develop prey mimicry through highly elaborated mantle tissue to attract a host fish. The often colorful conglutinate is released on to the stream bottom where it may resemble a prey item for the preferred host fish, such as a small fish, crustacean, worm, or insect. Prey mimicry by adult females has evolved in a number of genera, including *Lampsilis*. The function of the highly modified mantle margin is to cause an appropriate fish host to confuse the mantle margin with a prey species and then attack the gravid female mussel which then releases young into or on the confused host. The mantle margin may be striped and contain false eyespots to resemble a small fish, or it may look like another prey, such as a crayfish. The mussel *Epioblasma* has the highly unusual strategy of clamping its shell around the attacking fish (a darter in most cases) to hold it long enough to release large numbers of glochidia directly into the fish mouth and thence on to the gills.

III. ECOLOGY, BEHAVIOR, AND ENVIRONMENTAL BIOLOGY

A. Physical Habitat and Other Environmental Requirements

Mussels and clams are entirely benthic creatures except for the planktonic larval stage of zebra and quagga mussels. Unionid mussels are typically found in streams and rivers, but relatively few species inhabit lakes, ponds, and wetlands. Within these flowing water habitats, they spend almost all their time buried in mud, sand, or gravel, but avoid shifting sand and thick silty substrates. Thin-shelled species are more fragile and typically confined to soft sediments in slowly flowing environments, while thick-shelled species occur in all substrates and tolerate higher current velocities. Sphaeriid clams prefer sand to mud habitats in rivers to ephemeral ponds. Asian clams occupy perhaps the widest range of substrate types, including bedrock, but are most common in sandy-bottomed streams with intermediate flows. Zebra and quagga mussels live in lakes and rivers, where they attach to any available hard substrate with strong proteinaceous fibers. Unfortunately, the hard substrates include native unionid mussels, and this has resulted in mass mortality in mussel beds of these native species.

Although many factors influence the distribution and abundance of mussels and clams, the distribution and possibly individual growth rates are affected by the availability of calcium salts, especially at the extremes of concentration. Excessive suspended sediment can also limit mussels because it clogs the gills, thereby interfering with both respiration and feeding. And, of course, water pollution harms mussels and clams.

B. Trophic Position

Mussels and clams rely to varying degrees on food obtained from the water column (filter or suspension feeding) or surrounding benthic surfaces (pedal or deposit feeding), with the former apparently dominant for most species.

Bivalves obtain suspended algae and some other organic particles by filter feeding. To do so, gill cilia generate a current that pulls water in through the incurrent (inhalant) siphon, passes it over the gills and labial palps, and sends it out through the excurrent (exhalant) siphon. There is some controversy about the relative importance to food capture of the cilia, gill mucus, and water viscosity for small particles. Food captured by the gills are then sorted by the labial palps and other means, and either passed inward for digestion or packaged with mucus and expelled through the excurrent siphon as pseudofeces. Bivalves can filter an enormous amount of water each day. Unionid mussels can filter perhaps 1–2 L of water per hour per dried gram of mussel tissue, while Asian clams and dreissenid mussels can filter several times as much per hour. Scientists estimated that the total population of unionid and dreissenid mussels in Lake St. Clair (the small lake connecting Lake Huron and Lake Erie) can filter 1.5–5.3% of the total lake volume per day.

Another source of food is rich organic matter lying on or in the sediments which can be obtained by pedal feeding (i.e., using the foot) or through siphon-suction feeding (in some sphaeriids only). The relative importance of this food source is controversial and taxon-specific, but it seems clear that juveniles of many taxa and species with small adults (like sphaeriids) obtain some energy in this fashion, in part because their small gills are not as effective in filter feeding. This benthic feeding technique may also contribute to the diets of larger mussels. In some cases, ciliary tracts on the foot transport small (3–8 μm) particles to the gills, or food is swept into the mussel using the foot (= pedal sweep). As mussels grow in size, they tend to become less efficient at pedal feeding and must rely principally on suspension feeding. Asian clams seem adept at both feeding techniques, while adult dreissenid mussels are entirely filter feeders.

C. Biotic Interactions and Population Limiting Factors

Bivalve molluscs can be eaten by any predator that can either swallow the mollusc whole or gain access to the soft tissue by breaking the shell or entering through a damaged area. This makes small and thin-shelled species most vulnerable to predators while also providing a partial escape in size for many species. The latter accounts for common disparities in body size in mussel populations.

A variety of aquatic and terrestrial predators consume the soft tissue of mussels and clams. Turbellarian flatworms, like *Macrostomum*, can eat any mussel or clam small enough to be handled by its muscular pharynx, and crayfish can eat any bivalve whose shell is thin enough to break. Fish pick up mussels and clams and either swallow them whole or first crack their shells with oral or pharyngeal teeth. Some examples of fish predators are native catfishes, drum, pumpkinseed sunfish, sturgeon, whitefish, and the exotic black carp, which is an especially serious threat to our native bivalves. Aquatic vertebrate predators include a few frogs and salamanders (preying on sphaeriids) and turtles (mussels and Asian clams). Terrestrial predators include muskrats in particular but also other mammals (mink, otter, and raccoons) and birds (e.g., some crows, ducks, goldeneye, grackles, kites, limpkins, and scaups).

The health and reproduction of individual bivalves and the rate of population growth in mussels and clams are affected by many freshwater parasites and by bacterial (e.g., *Aeromonas*) and viral diseases (e.g., noroviruses). These are especially common in bivalves because the parasites can easily gain access to the mantle and visceral mass through incurrent and excurrent siphons. It is not always clear, however, how many of the observed "parasites" found in the mussel are really noninjurious ecto-commensals, but some of these are certainly parasitic and can reduce growth rates and cause sterility. Some examples of parasites and commensals are thigmotrich ciliated protozoa, enteric parasites like *Cryptosporidium* and *Giardia*, trematode flatworms, bryozoans, aquatic earthworms, leeches, and a few insects.

Competitive interactions in mussels and clams are difficult to document, but it is clear from field observations of Asian clams and zebra mussels that filtration by these exotic species can reduce phytoplankton abundance in some rivers and lakes by greater than 50% and increase water clarity by comparable or greater amounts. Reduction in such a major trophic component has widespread impacts that ramify throughout the food web from other algae, to herbivores like native mussels, and all the way up to fish and birds.

IV. COLLECTION AND CULTURING

Given the many threatened and endangered species of pearly mussels, we recommend against removing live mussels from their native habitat, especially if threatened species occur in your local area. In most cases, you can easily identify taxa with empty shells you find in the area. If you do remove a live mussel from a stream or pond, you should be able to identify it to the level used in this field guide in a few minutes and then return it to the water. To avoid potential errors in how it is inserted in the bottom, just lay the mussel in the area you found it and allow it to dig its own way into the bottom. If the mussel is to be displayed in an aquarium, it should be taken to the laboratory in a cooler on ice and later returned to the wild within a couple of days to avoid likely starvation for these mostly suspension-feeding molluscs. Permanent removal of the mussels from the field usually requires special permits.

If you are participating in a sponsored survey to determine environmental quality of a stream or pond, you can collect bivalves at least semi-quantitatively with linear transects, quadrats, or timed searches in a given area following the procedures outlined in your survey manual and preferably when the water level is low. The number of species collected and the mean size usually vary substantially between timed and quadrat searches.

A common method for collecting pearly mussels is to use hands or feet to feel along the bottom in shallow water for the shells on sandbars and stream banks. The closer to the shore, the more likely the outer shell has been extensively weathered especially in turbulent streams or wave-prone lakes. You can also use a viewing bucket in clear water or rakes to find and collect live mussels.

Commercial mussel harvesters sometimes use a brail (a metal bar with lines and hooks) to collect mussels in rivers or dive with scuba tanks or surface-supplied air, but these methods require special training and are subject to strict legal regulations.

Sphaeriid and invasive Asian clams are usually more abundant than pearly mussels but are harder to see on the bottom because of their size or tendency to burrow deeper. While mussels occur in both rocky and soft sediment habitats, freshwater clams are most commonly found in muddy sand to softer sediment habitats where they are easiest to collect with a shovel or professional grab sampler (e.g., Ekman or Ponar) from depths of 1 m or less (sometimes to 5 m). Areas with extensive weed beds are likely spots to find sphaeriids. After collecting the sediment, wash it through medium to fine seives and either first search the sieve for the clams or pour the material into a white tray for easier examination.

Juvenile Asian clams and larval dreissenids can be collected from the water column in rivers in certain seasons, but identification and counting of the latter require a dissecting microscope.

Feel free to collect all the exotic dreissenid mussels you want as long as you do not transport them to a new aquatic habitat. In almost all cases, the zebra and quagga mussels are found attached to each other and a hard surface (e.g., rocks, pipes, dead wood, boat hulls, or unfortunately living and dead pearly mussels) at depths of a half meter or so down to the depth that their phytoplankton food occurs. The living mussels can be separated from the hard surface most easily by cutting their fine byssal threads with a pocketknife. Be careful when handling them in large clumps because the shells can be very sharp.

Culturing bivalve molluscs is very challenging for the amateur because a steady supply of the proper kind of phytoplankton is required to keep them alive, and they are also sensitive to oxygen, thermal, and nutrient conditions in the water. They can, however, be brought into a classroom or home for observation over a period of a few days and then returned to the field. Scientists continue to develop techniques for culturing and propagating pearly mussels in an attempt to reintroduce them into areas where they have been extirpated or where populations are low.

V. REPRESENTATIVE TAXA OF CLAMS AND MUSSELS: CLASS BIVALVIA

A. Family Sphaeriidae

Fingernail, Pea, and Pill Clams: *Pisidium sp. (Fig. 10b), Musculum sp., Sphaerium sp., and Eupera sp.*

Identification: Small, free living clams, never attached. Color yellow to brown or orange, green, or blue, often with darker growth stripes across body; foot and mantle typically whitish. Growth lines rarely raised. *Pisidium* with hinge beaks shifted posteriorly. **Size:** 1–10 mm wide. **Range:** Numerous species, often co-occurring; found from arctic south throughout continent. **Habitat:** Common in large rivers and lakes to 3 m deep and in well-oxygenated creeks and ponds. They occur in sandy or gravelly bottoms in streams and in soft, vegetated substrates of lakes and ponds. **Remarks:** If held in your still hand in some water and in the shade, the clam may open and move over your hand with its foot.

Fingernail, Pea, and Pill Clams: *Pisidium sp., Musculum sp. (Fig. 10c), Sphaerium sp., and Eupera sp.*

Additional Remarks: Hinge beaks toward anterior of shell. Shell thin.

Fingernail, Pea, and Pill Clams: Pisidium sp., Musculum sp., Sphaerium sp. (Fig. 10d), and Eupera sp.

Additional Remarks: Hinge beaks toward anterior of shell. Shell thick, strong.

Fingernail, Pea, and Pill Clams: Pisidium sp., Musculum sp., Sphaerium sp., and Eupera sp. (Fig. 10e)

Additional Remarks: Hinge beaks toward anterior of shell. Limited to coastal drainages from South Carolina to Texas.

B. Family Corbiculidae

Asian clams (Fig. 10f): Corbicula sp.

Identification: Free living, never attached to anything. These are small- to medium-sized clams, ranging in color from yellow to brown, and sometimes with darker stripes. Growth lines are raised arcs across the shell. The foot and mantle are typically white to off white. **Size:** 1–40 mm in width. **Range:** Native to Asia, these clams have been introduced throughout the USA and southern Canada. **Habitat:** Common in large rivers, streams, creeks, ponds, and lakes to 3 m deep. These clams occur on almost any substrate from bare rock to gravel, sand, and mud. **Remarks:** Commonly sold in bait shops for bottom fish, sometimes sold in pet shops.

C. Superfamily Unionoidea

Pearly Mussels (Fig. 10g): Heelsplitters, Pigtoes, Floaters, Deertoes, Monkeyfaces, Mapleleafs, and others: Anodonta sp., Eliptio sp., Amblema sp., Gonidea sp., Margaritifera sp., Obovaria sp., and many others.

Identification: Free living, never attached. Color brown, black, gray, yellow, or orange, often with darker stripes, and white or yellow patches where the shell is worn; foot and mantle typically whitish. Growth lines never raised, but flange may project from one side or shell ribs may radiate from hinge line. **Size:** 1–18 cm. **Range:** Subarctic. Predominately east of Rocky Mountains, with nearly 300 species. Only nine species are found west of the Rocky Mountains. **Habitat:** Large rivers and streams and well-oxygenated lakes to a depth of 3 m in gravels, cobble, and sand. **Remarks:** Many species are protected by federal, state, or local laws or fishing ordinances. Check your local laws before handling these animals.

Pearly Mussels (Fig. 10h):

Additional Remarks: Example of variation in shell morphology.

Pearly Mussels (Fig. 10i):

Additional Remarks: Example of variation in shell morphology.

Pearly Mussels (Fig. 10j):

Additional Remarks: Example of variation in shell morphology.

Pearly Mussels (Fig. 10k):

Additional Remarks: Example of variation in shell morphology.

D. Family Dreissenidae (True Mussels)

Zebra Mussel (Fig. 10l): Dreissena polymorpha Pallas, 1771

Identification: Triangular mussels attached by threads to hard surfaces. Color white or yellow to brown; usually with brown striping. One side of shell flat; valves with single ridge line. Line joining the shell valves at posterior is straight. **Size:** 3–15 mm long. **Range:** Native to Caspian Sea region, but introduced widely in Europe, USA, and Canada (after entering through Great Lakes). **Habitat:** Rivers, streams, lakes, and ponds to a depth of 3 m or more. Typically found attached to hard surfaces such as rocks, pilings, docks, boats, litter, crayfish, and other molluscs. **Remarks:** A highly invasive species that can travel easily for long distances when attached to a boat, boots, or fishing gear.

Quagga Mussel (Fig. 10m): Dreissena bugensis Andrusov, 1897

Identification: Attached by threads to hard biotic and inanimate surfaces like zebra mussels. Triangular to oval mussels; color off white, yellow, or brown, occasionally with brown striping. Both sides of shell rounded and each valve lacking ridge lines. At posterior of shell, the line joining the valves curves. **Size:** 3–20 mm long. **Range:** Native to Caspian Sea region, but introduced to Europe and into Great Lakes and now spreading through Mississippi and Colorado Rivers and across continent. **Habitat:** Rivers, streams, lakes, and ponds; most common to 3 m. **Remarks:** This is a highly invasive species that can travel easily for long distances when attached to a boat, boots, or fishing gear.

False Dark Mussel (Fig. 10n): Mytilopsis leucophaeata (Conrad, 1831)

Identification: Attached by threads to a hard surface. Triangular, but less angular than the zebra mussel. Shell is brown, occasionally with brown striping. Both sides of shell rounded and each valve lacking ridge lines. Inside the shell near the hinge is a tooth-like projection that is absent in other freshwater true mussels. At the shell posterior, the line joining the shell valves curves. **Size:** 3–20 mm long. **Range:** Native to the Atlantic and Gulf Coast. **Habitat:** Estuaries and fresh brackish habitats to a depth of 3 m. Typically found attached to hard surfaces such as rocks, pilings, docks, and boats. **Remarks:** This species is a nuisance fouling invasive species in Europe and Africa.

Chapter 11

Aquatic Segmented Worms and Leeches: Phylum Annelida

I. INTRODUCTION TO THE PHYLUM ANNELIDA

If you have planted a garden in the backyard or waded in a weedy lake or wetlands in many parts of the USA, you have undoubtedly encountered segmented worms, such as terrestrial earthworms (including night crawlers) and leeches—all members of the phylum traditionally called Annelida. This chapter focuses on three groups of aquatic annelids: freshwater earthworms or oligochaetes (all of which lack suckers); leeches with their two suckers (anterior and posterior) or one posterior sucker; and branchiobdellidans (crayfish worms), which possess only a posterior sucker and live as commensals on the exterior of crayfish and a few other organisms. Two other groups of annelids colonize inland waters: about a dozen, mostly brackish water species of the predominately marine group Polychaeta, and the leech-like *Acanthobdella pelidina* (the only North American species), which is an ectoparasite on salmonid fishes. Because both groups are relatively rare in freshwaters, they are not discussed further in this chapter.

The classification of middle to higher taxonomic levels of this phylum (order to class and higher) is currently in a state of flux, but for this field guide we can ignore this controversy and focus on the general characteristics of the lower taxonomic levels. If you wish more information on this controversy and annelid biology and classification in general, please consult Govedich et al. (2010).

North America contains over 200 species of freshwater earthworms, 76 known species of leeches, and at least 107 species of branchiobdellidans. None of these are listed on state and federal lists of threatened or endangered species, but we suspect at least some of the cave species of branchiobdellidans are at risk because of threats to their host crayfish. Identification of most freshwater annelids is very difficult without a good microscope.

Freshwater annelids generally can be divided into two feeding groups—those that primarily consume dead organic matter (almost all oligochaetes), predators and parasites (leeches), and omnivores (branchiobdellidans). They primarily live in or on the bottom of lakes, wetlands, and mostly slow-flowing areas of creeks and rivers, or they colonize the surface of crayfish and some other invertebrates.

II. FORM AND FUNCTION

A. Anatomy and Physiology

The soft-bodied annelids are easily distinguished from other invertebrates by the combination of easily seen segmentation, the lack of major visible external differences among segments (except for the presence of suckers in leeches and branchiobellidans), an elongated and cylindrical (in cross section) body, and the absence of a head, visible mouthparts, appendages, eyespots, external

ISBN 978-0-12-381426-5, DOI: 10.1016/B978-0-12-381426-5.00011-9

genitalia, and antennae. Aquatic annelids normally respire aerobically using the skin surface; but under low-oxygen conditions, some can respire anaerobically, though the respiration rate falls substantially. A few species also have external, gill-like structures.

Most aquatic earthworms are 1–30 mm, but some giants reach 15 cm. They have 40–200 segments, all but the first of which contain four bundles of short to long hairs, or chaeta (in comparison to the abundant hairs found in polychaetes) in most species. Most are brown or transparent, but some species living in low-oxygen sediments are bright red because of the hemoglobin in their blood.

Most adult branchiobdellidans are barely large enough to see with the naked eye under good circumstances, but as a group they range in length from 0.8 to 10 mm and often have a whitish or more transparent color. The body is mostly rod- or spindle-shaped and somewhat flattened; their head is usually distinct from the body. They have a constant number of 15 segments. The eleventh segment is modified into a single posterior, disk-shaped sucker which is used to attach to the host). They move in a "leech-like" or "inch-worm" fashion using this rear sucker and anterior adhesive organ, both of which can produce chemical bonds that can be broken in one second.

Leeches are characterized by a nonsegmented prostomium (a section in front of the mouth), 32 segments behind the mouth, absence of chaetal hairs, an anterior (oral) sucker that is typically smaller than the posterior (anal) sucker, and a number of other exterior and interior features. All leeches use the "leech-like" locomotion involving alternate attachment and release of one and then the second sucker, but many species can swim by using dorsoventral undulations of the body. They vary in size from half a centimeter to nearly 5 cm in length in extreme cases. Unlike most other annelids, leeches are frequently brightly colored. The background body color is usually dark brown or black, but this is overlain by stripes, spots, and other patterns of bright red to yellow and even green colors.

B. Reproduction and Life History

Annelids are hermaphroditic organisms with either males produced first (protandry) or both sexes present simultaneously. When they mate, the female stores sperm and eggs. In most species the female then secretes a ringed cocoon around the body, which eventually slides over the gonopore (picking up fertilized eggs) and the head before it is deposited on a suitable rock or other firm substrate, including a host crayfish in the case of branchiobdellidans. A few species of worms are parthenogenetic. Aquatic annelids seem to live a maximum of 1 or 2 years and generally breed only once, with leeches living the longest. Blood-feeding leeches, however, more commonly breed multiple times, with three or more blood meals required to attain sexual maturity. Some leeches brood their young and may provide some external nutrition through their body wall.

III. ECOLOGY, BEHAVIOR, AND ENVIRONMENTAL BIOLOGY

A. Habitats and Environmental Limits

Aquatic earthworms occur in permanent and many ephemeral water bodies with some organic sediments—which typically means lower or zero current velocities in the microhabitat. Most species are cosmopolitan, or at least widely distributed. Other species have more limited geographic distribution and may be confined to a single water body, such as Lake Tahoe or the Great Lakes or to certain cave systems in the eastern USA. They are relatively tolerant of pollution in general, and the abundance of some species is a biotic indicator of generally poor water quality. In environments with low-oxygen conditions and fewer predators and competitors, their densities can reach 8000 or more per square meter.

Branchiobdellidans are almost entirely obligate commensals on crayfish, but it is probably wrong to characterize them as parasites. Some species have been reported living independently of crayfish, but all species seem to require their cocoons to be attached to a host for proper development. The number of branchiobellidans is quite variable, but in one case 1800 individuals from five genera were found on a single crayfish host. Crayfish are the most common host, with most genera present on only one side of the Rocky Mountains. South of the distribution of these crayfish, branchiobdellidans typically colonize freshwater shrimp and crabs. In North America, these annelids also live on isopods in caves. Crayfish have been observed attempting to remove the annelids from their gills and body, but this is rarely very effective for long.

Leeches are benthic species in general (living around many types of substrates), but are commonly found above the bottom and will occasionally swim in search of prey, hosts, or better habitats. They are more abundant in shallow waters (<2 m), but some inhabit very deep waters. Five hundred or more may coexist per square meter in suitable habitats. They occur in a wide range of salinities from very soft waters to habitats with salt concentrations greater than seawater. Although most are found in well-oxygenated waters, many species can exist in anoxic conditions for days or weeks.

B. Functional Roles in the Ecosystem and Biotic Interactions

Freshwater annelids are generally detritivore, predators, or omnivores, with prominent differences among the large taxonomic groups discussed here. Oligochaetes ingest sediments and extract the nutritious bacteria as well as any algae or microinvertebrates associated with it. Some species feed more on the bottom surface of the water body where they can obtain relatively more algae and less dead organic matter, but most live in the upper 5–10 cm of the bottom and continually process the organic matter therein. Several genera of aquatic earthworms in North America are primarily predators on various small invertebrates, and some groups prefer eating the bacterially colonized feces of other worm species. The vast majority of branchiobdellidans are opportunistic omnivores which will eat whatever they can find present externally on their crayfish host (small invertebrates, algae, dead organic matter) or that is inadvertently released when their host is feeding. A few species are considered primarily carnivores, and most will apparently ingest host tissue exposed by an inadvertent break in the outer cuticular covering. Some scientists feel that these annelids will consume some host tissue, like gills, on occasion (thus making them ectoparasites), but this has not been adequately proven. Leeches are best known to people as bloodsuckers, and indeed many species are temporary parasites (dropping off after their blood meal) on fish, turtles, alligators, amphibians, water birds, and an occasional human and crayfish. However, most leech species prey on midges, earthworms, scuds, and snails. The young of several leech species attack zooplankton, including the leech *Motobdella montezuma* which preys principally on planktonic amphipods in its unique and ancient natural well. The community-wide effects of leech predation are unknown, but attacks on many invertebrates species are fatal. Phylogenetic evidence suggests that leeches evolved from sanguivorous ancestors but lost this "blood-feeding" habit at multiple, independent times in their species radiation. Sanguivorous leeches produce anticoagulants to aid flow of their victim's blood; their properties are being investigated for human pharmacological applications.

Predators of annelids are abundant and diverse. Benthic feeding fish and many species of invertebrates consume aquatic earthworms. Most oligochaetes have tails that can mechanically sense the approach of potential predators, causing the worm to withdraw rapidly within the sediments. Leeches are attacked by fish, birds, amphibians, and various water snakes as well as by some insects and crustaceans. Branchiobdellidans are subject to the same predators that attack their host crayfish. In addition, ectoparasitic ciliate protozoa and rotifers feed on them, though usually not fatally.

IV. COLLECTION AND CULTURING

Collection techniques for freshwater annelids vary among the three major groups. Collect aquatic earthworms with any grab sample from the organic-rich bottom of ponds or streams, and then sort the sample in a shallow white pan. The worms will usually thrash weakly about the bottom or become tangled in a knot. Branchiobdellidans can be collected by examining live crayfish in the field in a bowl with a hand lens. You can also take the live crayfish into your home or lab and gently scrape the annelids off the host if you want to examine them under a microscope. Leeches, while often abundant, can be more difficult to find. However, they can sometimes be collected by hand, with a net, or a grab sampler from hosts or various firm substrates, or while they are swimming through the water. They are often mostly easily collected in permanent wetlands, especially those without predaceous fish.

Rearing techniques also vary among the groups. Oligochaetes can easily be kept in an aerated aquarium with organic sediment. Replace the water partially every few weeks with filtered water from the natural habitat. They can be fed with sinkable fish food or formerly frozen lettuce buried in the bottom. Branchiobdellidans are best maintained in an aquarium on live crayfish, but you can occasionally maintain them in a small bowl filled with water from the pond or stream and kept at relatively cool temperatures of 10–15°C. Raise leeches in an aquarium under approximately constant conditions of temperature, salinity, and oxygen. Feed non-sanguivorous leeches weekly by providing abundant, live invertebrate prey. Bloodsucking leeches are more difficult for the amateur to maintain in the home or lab, but techniques are available (e.g., providing them warmed blood encased in the epithelial skin used to make sausages).

V. REPRESENTATIVE TAXA OF AQUATIC EARTHWORMS AND LEECHES: PHYLUM ANNELIDA

A. Aquatic Earthworms (Fig. 11a): Class Oligochaeta

Identification: Segmented worms with more than 15 segments. Round in cross section. Free living, without suckers, ranging in color from white, pink, red, yellow to brown or orange, green, or blue. Sometimes the gut is visible as a red stripe through the body. **Size:** 1–30 mm, rarely to 15 cm. **Range:** Numerous species, with several sometimes occurring together, ranging all over North America. **Habitat:** Common in rivers, streams, creeks, springs, ponds, wetlands, and lakes to 10 m deep. Usually in soft substrates such as mud, silt, or sand. **Remarks:** In poorly oxygenated water, members of the genus *Tubifex* will wave one end of their bodies in the open water in an attempt to gather oxygen.

B. Polychaete Worms: Class Polychaeta

Hobsonia florida (Fig. 11b)

Identification: Segmented worms with a ring of tentacles at the head. Round in cross section. Free living, without suckers, ranging in color from white, to pink, or red. **Size:** 1–10 mm long. **Range:** Atlantic coastal USA from New England, south, across the Gulf Coast states. Possibly introduced to coastal Oregon. **Habitat:** Estuaries to 3 m deep. In soft muddy substrates. **Remarks:** This is a filter feeding tube worm.

Manayunkia sp. (Fig. 11c)

Identification: Segmented worms with a ring of oral tentacles. Round in cross section. Free living typically in tubes. Suckers absent. Animals are white, gray, pink, or red in life. **Size:** 1–5 mm. **Range:** Great Lakes Region east through Ontario, Quebec and New England, and in mid-Atlantic coastal estuaries. Introduced to Alaska, British Columbia, Washington, Oregon, and California. **Habitat:** Estuaries to 3 m deep. In soft muddy substrates, high in organic material. **Remarks:** Two species are known from fresh and brackish waters: *Manayunkia aestuarina* an estuarine species and *Manayunkia speciosa* from freshwaters.

C. Crayfish Worms (Fig. 11d): Class Branchiobdellida

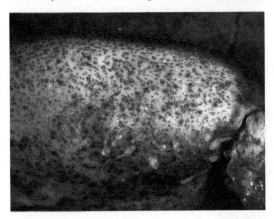

Identification: Segmented worms without tentacles, no oral sucker, posterior sucker present. Round in cross section, sometimes with dorsal projections. Color in life may be from white, gray, brown, or black. **Size:** 0.8–10 mm. **Range:** Found throughout North America, wherever the hosts are present. **Habitat:** Habitat is dependent on hosts. **Remarks:** Branchiobdellans live externally on crayfish, shrimp, and crabs, feeding on other animals and periphyton that grows on the host, or sharing the food of the host.

D. Leeches: Class Hirudinea

Fish Leeches (Fig. 11e): Family Piscicolidae

Identification: Segmented worms with one oral and one posterior sucker, and no tentacles. Round in cross section. Color of living animal may be from white, gray, brown, green, orange, red, yellow, or any combination of these colors. **Size:** 0.8–15 mm. **Range:** Found throughout North America, wherever the hosts are present. **Habitat:** Fresh and brackish water. Ponds, lakes, rivers, and streams. Often found on the host. **Remarks:** Fish leeches are almost invariably parasites on fish.

Leeches (Fig. 11f): Family Glossiphonidae

Identification: Segmented worms with one oral and one posterior sucker, and no tentacles. Broadly flattened in cross section. Color of living animal may be from white, gray, brown, green, yellow, or any combination of these colors. **Size:** 0.8–20 mm. **Range:** Found throughout North America. **Habitat:** Ponds, lakes, rivers, streams. Under rocks or in vegetation. **Remarks:** These leeches are not parasites, but instead are predators. Females may carry the young underneath their body.

Leeches (Fig. 11g): Family Erpobdellidae

Identification: Segmented worms with one oral and one posterior sucker, and no tentacles. Round to slightly flattened in cross section. Color of living animal may be from white, gray, brown, green, yellow, or black. **Size:** 0.8–70 mm. **Range:** Found throughout North America. **Habitat:** Ponds, lakes, rivers, streams. Under rocks or in vegetation. **Remarks:** These leeches are predators on snails, crustaceans, worms, and insects.

Chapter 12

Moss Animals: Phylum Ectoprocta, or Bryozoa

I. INTRODUCTION, DIVERSITY, AND DISTRIBUTION

Moss animals are widely distributed but a rarely noticed group of colonial animals in the mostly marine phylum Ectoprocta. Their evolutionary origin is unclear, but they seem related to members of the small marine phylum Phoronida, which use ciliated tentacles to obtain food like bryozoans.

These microscopic creatures living in macroscopic colonies are commonly encountered in lakes and streams but are usually not recognized by nonscientists as animals, which is perhaps not surprising given their old common name "moss animals." Instead, colonies of bryozoans resemble anything from amorphous strings to massive gelatinous mounds weighing several kilograms. These benthic suspension-feeders are most easily found during warmer months attached to rocks, submerged branches, or even to the underside of floating docks.

A bryozoan colony consists of individual animals connected internally by a fluid-filled cavity and externally by a secreted nonliving protective coating. A prominent feature of moss animals is their often U-shaped lophophore (Fig. 12a). This structure contains ciliated tentacles, which are employed to filter organic matter from surrounding water.

Only about 24 bryozoan species have been identified from North America, all collected from east of the Mississippi River and north of the 39th latitude (about the level of Washington, DC); most of these are found around the Great Lakes states. Only one species, the brackish water *Victorella pavida*, is restricted to more southern latitudes. Scientists have published little about populations of bryozoans west of either the Mississippi River or the province of Ontario, and few agencies have recorded their presence. However, species in several genera are known from Arizona, California, Colorado, Nevada, Oregon, Utah, Washington, and British Columbia. About a quarter of the known species are considered rare.

Despite their general northern distribution, bryozoans are most commonly found in warmer months in areas where water temperatures reach around 15–28°C. Despite such preferences, some species can be collected in winter when temperatures are in the single digits and a few taxa in tropical countries survive temperatures of at least 37°C.

Some bryozoans were formerly nuisance species because they fouled intake pipes of water distribution systems. This problem largely declined when many utility companies started drawing water through sand filters, and now only some untreated intake systems face periodic difficulties in warmer months. Moreover, this fouling problem is now considered minor compared to the havoc wreaked by invasive clams and mussels (see Chapter 10).

Bryozoans superficially resemble members of the unrelated phylum Entoprocta (note the difference in spelling of the first syllable), with whom they have been erroneously classified at various times in the past. The confusion arose because both possess ciliated tentacles and an incomplete separation of budded units (zooids); however, the resemblance ends there. The latter phylum contains perhaps 60 species worldwide, but the only entoproct bryozoan in freshwaters of North America is

© 2011 Elsevier Inc. All rights reserved.
ISBN 978-0-12-381426-5, DOI: 10.1016/B978-0-12-381426-5.00012-0

FIG. 12A. Anatomy of a bryozoan colony.

Urnatella gracilis. Like other entoprocts, it has an external segmented stalk which supports the body. For the remainder of this chapter, however, all references to bryozoans pertain to the more diverse ectoprocts.

II. FORM AND FUNCTION

A. Anatomy and Physiology

A bryozoan colony is composed of identical linked animal units (zooids). An individual zooid consists of a partially protrusible organ system (= polypide) surrounded by a body wall. The polypide bears a large retractable lophophore with cilated tentacles used to collect food. Depending on the family, the lophophore may be a simple structure surrounding the mouth, or be elaborated as a small conical structure, or be developed as a large U-shaped structure of paired arms with outer long tentacles and inner short tentacles. These tentacles have rows of cilia that produce coordinated waves (metachronal) to move food to the mouth where a heavily ciliated structure (epistome) appears to select food items. Depending on the order, colonies may be compact and globular or diffuse and formed into delicate zooid chains. Ectoproct bryozoans have a peritoneum-lined body cavity (coelom) in the lophophore and gastric area, which links them to the other, more complex coelomate invertebrates and vertebrates. These sessile animals lack a head.

Individual zooids within the bryozoan colony lack neural links but share a common coelom where gametes and nutrients can potentially be exchanged. Connecting the zooids externally is a nonliving layer, the ectocyst. In tubular colonies, the ectocyst varies from thin and flexible to leathery and brittle; it is typically sclerotized and may contain chitinous microfibrils. In non-tubular bryozoans, the ectocyst is gelatinous and may be up to 99% water, as in colonies of the massive *Pectinatella magnifica*.

B. Reproduction and Life History

Depending on the group, bryozoans regularly reproduce asexually with encapsulated dormant buds called statoblasts, asexually with both fission and statoblasts, and sexually with larvae. In addition, all bryozoans produce new zooids by asexual budding. Statoblasts, which are produced by some bryozoans, are useful in species identification. Most contain gas for buoyancy (floatoblasts), but some are cemented to the substrate on which the colony is growing (sessoblasts) or are held within the colony's tubular branches (piptoblasts). Statoblasts can survive freezing and desiccation for a year or so and possibly as long as a decade. When the dormancy period is over, a single zooid is formed, which then produces 1–5 new zooids to begin the colony. These resistant, reproductive structures are replaced in some bryozoans by thick-walled hibernacula attached to the substrate which are resistant to desiccation and other stressful environmental conditions. Asexual fission is limited to the mostly globular bryozoans. Following fission, the new colonies separate by gliding in different directions across the substrate under the propulsion of beating cilia of the lophophores.

While asexual reproduction can occur at any time when the colony is formed, sexual reproduction is limited to a single and brief period per generation for temperate zone species. Fertilization occurs within and between colonies for at least one species, but the mechanism is known only for the former. Sperm are released into the common body cavity of the colony where they circulate and eventually fertilize clusters of ova. The fertilized embryo develops into a free-swimming form called a "larva" but which is in fact a small motile colony composed of four or fewer zooids. The larva is released at night and normally settles within an hour.

In most of North America, bryozoans produce at least two generations per year between spring and fall when water temperatures are at least 8°C. During the remaining time, the bryozoan exists as a statoblast or a hibernaculum. Exceptions are *Fredericella indica*, which commonly survives year-round even in water covered by ice, and *Plumatella repens* which experiences nearly constant environmental conditions in an Arizonan cave.

III. ECOLOGY, BEHAVIOR, AND ENVIRONMENTAL BIOLOGY

Most bryozoan species occur in both still and running water at temperatures under 30°C. They are more common in alkaline waters, but some can tolerate acidic pH conditions of 4.9–6.3. Modest turbidity conditions are tolerated by most species. Bryozoans are often considered indicators of clean water, but some species tolerate eutrophic conditions and organically polluted waters. Substrate availability may limit some populations even though they can thrive on a large variety of natural and human made material from rocks to aged wood to plastics and rubber. Some living aquatic weeds are also colonized, such as the underside of some species of large water lilies.

Bryozoans capture various food particles with their tentacles, including microorganisms (bacteria, algae, protozoa, small nematodes, rotifers, and small crustaceans), but the primary assimilated food source is probably small (<5 μm diameter) bacteria and fungi attached to dead organic matter in suspension. The contribution of feces and pseudofeces to the bottom of a lake inhabited by substantial populations of bryozoans can exceed that contributed by fish and waterfowl but is less than produced by the more efficient filtering mussels.

Information on predators of bryozoans is often anecdotal or at least indirect, and predation does not seem to be a major limiting factor for bryozoan populations. Some fish have been found with guts packed with bryozoans, but this may have been incidental and related more to predation on the midges living on the bryozoan. Some bryozoan tissue is toxic to fish, and crustaceans seem to eat them only when other food is scarce. Nonetheless, suspected predators, or parasites, or at least incidental pests include fish, crayfish, apple snails, caddisflies, midges, elmid beetles, mites, and flatworms. More important than these macro-predators, however, could be protozoans like myxozoan and microsporidian parasites. One myxozoan parasite of bryozoans, *Tetracapsuloides bryozalmonae*, is the most serious pest plaguing salmonid fisheries, and bryozoans may be the original, and less impacted host.

Bryozoans are known to be competitors for substrates with zebra mussels and may also compete with benthic algal species for space.

IV. COLLECTION AND CULTURING

You can collect bryozoans in shallow water on firm substrates submerged or floating in lakes and streams. A fruitful site to find gelatinous colonies or individual free statoblasts is on floating objects (including swimming or boat docks) at the waterline or slightly below. You can see most species easily with the naked eye, but a magnifying glass will help you spot statoblasts and confirm their identity at higher taxonomic levels. Identification of bryozoans to lower taxonomic levels requires examination of individual zooids and usually the statoblasts with a dissecting microscope.

Culturing bryozoans in the laboratory or home at least through the current generation is possible as long as suitable food is available. They should be collected from the field still attached to the substrate on which they were growing if possible, and the colony should be suspended upside down in an aquarium so that fecal matter and other debris will not settle on and foul the zooids. Maintain temperatures at or slightly below room temperatures (lower temperatures tend to promote survival but reduce colony growth rates). A variety of food sources have been tried with different success rates for various bryozoan species. These include pure algal cultures, a mixture of algal species and fine organic detritus from soil or pond, and even finely ground fish food from the pet store. Another option if space is available is to provide a steady supply of suspended organic detritus by circulating unfiltered water from a lighted aquarium with algae and fish, through a dark fish-rearing tank (causing the algae to die and decompose), and into the bryozoan tank.

V. REPRESENTATIVE TAXA OF MOSS ANIMALS OR BRYOZOANS: PHYLUM ECTOPROCTA

Branching Moss Animals (Fig. 12b): Plumatella sp., Fredericella sp., Victorella sp., Pottsiella sp., Stolella sp., Hyalinella sp., Cristatella sp.

Identification: Brown, yellow to gray and white or green, branching, linear colonies of zooids. **Size:** Up to 10 cm. **Range:** Subarctic regions of the world; common throughout North America. **Habitat:** Permanent lakes and ponds, rivers, and streams. Typically found on the underside of rocks or overhanging banks. **Remarks:** Often mistaken for algae or corals (which do not occur in freshwater).

Gelatinous Bryozoans (Fig. 12c): Pectinatella sp., Lophopus sp., and Lophopodella sp.

Identification: Gray or white to translucent, globular colonies of zooids; sometimes marked with red or green. **Size:** Up to 50 cm. **Range:** Subarctic regions of the world; uncommon in western USA and Canada. **Habitat:** Permanent lakes and ponds, sometimes rivers and streams. Typically found on the underside of rocks or overhanging banks. **Remarks:** Sometimes mistaken for algae or rotting flesh.

Chapter 13

Introduction to Freshwater Invertebrates in the Phylum Arthropoda

I. INTRODUCTION TO THE PHYLUM

Members of the phylum Arthropoda are found in all habitats on earth from high in the atmosphere (floating spiders) down to the deepest ocean trenches (many crabs and other crustaceans), and they represent more than 80% of the described species on earth. Estimates of arthropod diversity generally vary between 2 and 6 million species. While everyone is familiar with common terrestrial arthropods such as insects and spiders and many people regularly consume marine arthropods such as shrimp and crabs, arthropods are equally important in all inland waters from ephemeral wetlands to deep lakes and from headwater streams to great rivers.

This phylum is divided into five subphyla, one of which is extinct (Trilobita) and another of which lacks freshwater species in North America (centipedes and millipedes in the subphylum Myriapoda). The remaining three subphyla contain important freshwater representatives: freshwater mites and spiders in Chelicerata, aquatic insects in Hexapoda, and many types of Crustacea (e.g., water fleas and crayfish). Members of this phylum are distinguished from other invertebrates by their jointed appendages. These have evolved for many functions, including locomotion, defense, reproduction, food handling, and sensory reception. Arthropods are also successful because of the protection and support provided by their hard exoskeleton. To allow for continued growth, the exoskeleton is shed (molted) periodically during a process called ecdysis, a relatively strenuous and often dangerous endeavor. Their bodies are segmented, but these are frequently fused into body regions (tagmata). Aquatic mites have two tagmata, insects have three, and crustaceans have three, except where the head (cephalon) and thorax are fused into a cephalothorax. They range in size from microscopic copepods less than a tenth of a millimeter in length to the largest known freshwater invertebrate, the freshwater crayfish *Astacopsis gouldi* of Tasmania weighing 2–3 kg and measuring 40 cm long.

Arthropods were previously linked evolutionarily with the annelids, but the preponderance of molecular evidence now links them with other phyla that shed their outer skin in order to grow. This evidence suggests the closest links with the following phyla: Tardigrada (microscopic water bears), Onychophora (terrestrial velvet worms), Nematoda (free-living and parasitic roundworms), and Nematomorpha (hairworms, see Chapter 8).

A. Subphylum Chelicerata, Class Arachnida

Spiders, scorpions, and mites/ticks are the best-known arachnids to the average person, but only about 5000 or so species of freshwater mites (Acari) and a hand-full of quasi-aquatic spiders

ISBN 978-0-12-381426-5, DOI: 10.1016/B978-0-12-381426-5.00013-2

(Araneae) have been identified around the world out of the 98,000 described species in the phylum. The actual diversity of arachnids is poorly known because there are relatively few mite taxonomists worldwide despite the prominent role played by these arachnids in all ecosystems other than marine. Some scientists believe that the class's actual diversity may rival or surpass the number of insect species. The vast majority of freshwater mites belong to the Hydrachnidiae, but several other suborders have invaded freshwater from land independent of the main group of mites. Among the spiders, the most highly adapted to aquatic environments of temperate North America is the genus *Dolomedes*. Arachnids can be distinguished from other freshwater arthropods by their distinctive mouthparts (paired chelicerae and paired pedipalps), four pairs of walking legs, and no antennae. A thorough search of a square meter of bottom in a wetland, lake, or stream can reveal as many as 2000–5000 mites representing 50–75 species. Freshwater mites tend to be ectoparasites or predators, depending on the species and life stage. For more extensive information on this group see Chapter 14.

B. Subphylum Hexapoda, Class Insecta

A strong minority of the one million or so described species of insects (Insecta) and collembolans (Entognatha) live in freshwater habitats, but these are distributed among 10 orders, with half the orders being almost entirely aquatic, at least as larvae. The most aquatic orders contain insects such as mayflies, hellgrammites, dragonflies, stoneflies, and caddisflies. The remaining five orders have a minority of aquatic species, but nonetheless can be important members of aquatic communities; these include true bugs, beetles, midges, aquatic moths, and spongillaflies. Their life cycles involve either a sequence of egg, larva/nymph, and adult (in the orders Ephemeroptera, Odonata, and Plecoptera) or a succession of egg, larva/nymph, pupa, and adult. Something over 7200 species have thus far been identified in North America. The insect body consists of three tagmata: head, thorax, and abdomen. The head has a pair of antennae, a pair of compound eyes, up to three simple eyes, and a complex set of paired mouthparts. The thorax contains most of the organ systems; adults may also have three pairs of five-part legs and wings or wingpads in some taxa. The abdomen may have up to 10 visible segments and contain gills and lateral and/or terminal, non-locomotory appendages.

This subphylum also includes the small group of semiaquatic springtails (class Entognatha, order Collembola) which were once classified as insects. These are small (usually <6 mm long), wingless arthropods with a distinct head with paired antennae, a three-segment thorax with three pairs of legs, and a six-segment abdomen. Young springtails resemble adults except for their size and lack of reproductive organs. Collembola can be readily recognized by the furcula, a midventral abdominal appendage, which is normally folded forward under the abdomen. When released, it springs backward, propelling the "springtail" several centimeters through the air. A common aquatic habitat for the perhaps 10 semiaquatic species of springtails in North America is the water surface where they feed primarily on algae, detritus, and possibly bacteria.

Chapters 19–27 contain more detailed information on hexapods, including the many orders of insects.

C. Subphylum Crustacea, Classes Branchiopoda, Maxillopoda, Ostracoda, and Malacostraca

Crustaceans are extremely important members of the planktonic and benthic communities in almost all inland water ecosystems, including wetlands, freshwater and saline lakes, caves, and rivers, even though they represent worldwide only about 15% of the nearly 68,000 described species in the subphylum (with almost all the remaining being marine species). About 1500 freshwater species have been described from North America. Included among the inland water representatives of this subphylum are water fleas, tadpole shrimp (Fig. 13a), fairy shrimp (branchiopods), copepods, seed shrimp (ostracods), scuds (amphipods), water sow bugs (isopods), shrimp, crayfish, and crabs. They are a vital food web link between primary producers (algae and aquatic weeds) and higher trophic levels. Crustaceans obtain food as detritivores, herbivores, omnivores, and carnivores, and they range in size from minute zooplankton only 0.1 mm long up to giant crayfish at 400 mm.

FIG. 13A. Tadpole shrimp (Crustacea, Branchiopoda).

While members of the other subphyla have fairly standardized body forms, crustaceans are highly diverse, primarily because of variable fusion of body segments and development of very specialized appendages. Their jointed, biramous appendages may present in all three body regions. Unlike insects, adult crustaceans have two pairs of antennae. Their chitinous cuticle is often elaborated as a shield-like carapace, and their abdomen typically has more than 11 segments.

More detailed information on crustaceans can be found in Chapters 15–18,

II. FORM AND FUNCTION

A. Anatomy and Physiology

Arthropods have the most complex internal anatomy of all invertebrates other than in squids and other molluscan cephalopods. They are a coelomate phylum with an open circulatory system, which transports oxygen obtained by gills, tracheae, book lungs, and/or surface respiration. Wastes are excreted in aquatic species mostly as ammonia. Arthropods have a highly developed neural system with a cephalic brain and complex sensory organs, including image-forming eyes in some species. The most highly developed arthropod brain is more complex than those present in any other living organism other than vertebrates and cephalopods. Consequently, arthropods often have a relatively sophisticated behavioral repertoire and communicate visually, tactilely, and chemically.

The chitinous- and protein-based exoskeleton is often fortified with calcium carbonate and provides support, protection, and diverse locomotory options (walking, burrowing, swimming, and flying, when modified appropriately). Internal projections of the exoskeleton allow muscular attachments for this locomotion. The exoskeleton also reduces water loss for terrestrial stages and provides minor to significant protection from predators. Unfortunately, this useful external skeleton came with a cost associated with growth. To expand the body size or modify structures like appendages, the exoskeleton must be shed periodically in a molting process termed ecdysis. This process demands considerable energy to both form the new exoskeleton (especially if the old one is not later consumed) and absorb needed salts to strengthen it. Moreover, feeding may cease for a time before and after ecdysis. The physical process of shedding the old skin can be dangerous even if a predator is not lurking close by to snap up the temporarily soft shell animal because extracting the body from the old exoskeleton can be strenuous and even result in death.

B. Reproduction and Life History

Most freshwater arthropods are dioecious (two sexes) but hermaphroditism and even parthenogenesis occur. Courtship and brood care are uncommon in all subphyla of arthropods and are usually simple when they present. All freshwater arthropods have aquatic or semiaquatic larval stages. Most are short-lived compared to adult stages, but the opposite is true for most insects. Crustaceans typically undergo larval development within the egg/embryo rather than as a free-living stage, and even pre-adult stages may seem very similar to the adult except in size and elaboration of some structures (e.g., young crayfish while attached to the mother's abdomen). However, the nauplius larval stages of branchiopods and copepods differ dramatically from the adult forms in structure and size.

III. ECOLOGY, BEHAVIOR, AND ENVIRONMENTAL BIOLOGY

It is challenging to highlight aspects of the ecology of freshwater arthropods because they live in almost all habitats and fit within all trophic levels other than producers. Some exceptions to this generalization can be found among communities in planktonic, saline lake, thermal, and cave habitats. With minor exceptions, the only relatively permanent crustacean members of open water plankton communities (= holoplankton) are copepods and cladocerans. The remaining taxa are primarily benthic or in relatively few cases live at the water surface (neuston) in calm areas of ponds and streams. Examples of the latter are predatory water striders (Hemiptera, Gerridae) which skate across the water surface. Some insects and mites enter the plankton only as early instars (meroplankton) or live in the water column in association with aquatic weeds (littoral zooplankton). Phantom midges (Diptera, Chaoboridae), however, spend their entire aquatic stage migrating from near the bottom during the day up to near the surface each evening. Saline lakes are inhospitable to most arthropods, but brine shrimp (Branchiopoda, *Artemia*; marketed commercially as "sea monkeys") thrive at salinities much higher than those in the open ocean, as do brine flies (Diptera, Ephydridae). Very few invertebrates can thrive in hot spring habitats, but an arthropod exception is the shore fly *Ephydra brucei*, which can tolerate temperatures of at least 44°C. Another fly in the family Ephydridae, *Helaeomyia petrolei*, is the only known insect whose larva colonizes natural habitats of crude petroleum. Subterranean aquatic habitats support low populations of permanent freshwater residents because of low food supplies (all of it originates from outside the cave) and the overwhelming challenges for species having a terrestrial flying stage. The most common aquatic arthropod residents of caves are crustaceans, especially scuds, water sow bugs, and crayfish.

Although arthropods occupy most trophic levels and functional feeding groups in freshwater ecosystems, parasitism is relatively rare. Two exceptions are fish lice (class Maxillopoda, subclass Branchiura) and many mites. Fish lice are represented in North America by the genus *Argulus*. These crustaceans use head appendages modified as toothed hooks or suckers which enable the larva and adult to latch on to the gills, mouth, or skin of a passing fish from whom they extract blood meals or feast on extracellular fluids or mucus. Many mites are ectoparasites at some life stage and will select bivalve mollusc, insects (e.g., dragonflies), or other invertebrates as hosts.

Most freshwater arthropods have life spans equal to, or less than a year. In some cases, a generation may last a few weeks at best (e.g., crustacean zooplankton and chironomid midges), with multiple generations (cohorts) produced under favorable environmental conditions. Other taxa survive most of a year, and a few live 2–3 years. Among insects, northern temperate zone species tend to live longer than those in warmer southern latitudes. This may relate to the time necessary to accumulate enough energy for reproduction. A prominent exception to the short arthropod life is the generally long-lived cave crayfish. Some estimates suggest a life span of a century for some species, with maturity coming only after several decades. Even if troglobitic crayfish live only a few decades, their life span considerably exceeds that of other North American arthropods, including other crayfish. Constant environmental conditions and a shortage of food are probably the main factors contributing to their longevity. The delayed reproductive period and reduced fecundity also, unfortunately, make them especially vulnerable to extinction from groundwater pollution or other disturbances from humans.

Almost all freshwater invertebrates spend most of their lives in aquatic habitats, not counting any resistant eggs (really embryos) that are produced to survive dry conditions in floodplains or ephemeral pools. Most freshwater insects have a winged life stage, which is either terrestrial or semiaquatic. The exceptions are many bugs (Heteroptera) and a few beetles and stoneflies. A terrestrial adult reproductive stage tends to be short in comparison to the larval stage. The terrestrial phase may last a day and not include a feeding period, as in many mayflies, or as long as a month or more in dragonflies and feature active energy accumulation. After mating, eggs are laid in wetlands, ponds, or streams. Other semiaquatic species, such as beetles and true bugs, require atmospheric oxygen as adults but feed on aquatic organisms and live on or near the surface. Many mites are aquatic throughout their lives, while others may be ectoparasites on aquatic insects during their terrestrial phase; for example, colorful mites can often be found on adult dragonflies. Most crustaceans spend their entire lives in water or at least in a humid burrow on land except for brief periods of migration across land by some crayfish and crabs.

IV. COLLECTION AND CULTURING

Information on collecting and culturing arthropods is described in subsequent chapters. In general, if you use a plankton net, dip net, or collect by hand, you are certain to collect at least one arthropod, though it may require a microscope to spot those in plankton samples. The sole exception is a sample taken from a more demanding environment, such as a cave or hot spring, where more diligent sampling efforts may be required. If you preserve a grab samples of organic matter with hidden small organisms in ethyl or isopropyl (rubbing) alcohol, adding a few grains of any dye that stains tissue (e.g., Rose Bengal or Phloxine B), as this will make small arthropods more visible when you later pick through the sample.

With a limited exceptions (e.g., some dragonflies), few arthropods other than crustaceans appear on lists of threatened and endangered species. The most seriously threatened species are often crayfish in caves or many rivers as well as fairy shrimp and tadpole shrimp in ephemeral wetlands. In general, avoid taking animals from these environments. Also, consult Chapter 3 about general sampling and permits that may be required.

Many crustaceans can be reared through multiple generations in the lab, while most arachnids and insects can only be kept for short periods because of reproductive requirements, such as a terrestrial adult stage or the need to find a parasitic host. If you keep crayfish in the lab, be aware that they are adept at escaping most aquaria, especially if they can crawl out around a water filtration pipe.

V. TAXONOMIC KEY TO THE PHYLUM ARTHROPODA

The following taxonomic key is meant to take you to the appropriate chapter for members of the phylum Arthropoda. Included are one or two figures illustrating some of the arthropods representative of that couplet in the key, but many more examples are shown in Chapters 14–27 for each group. Some taxa in these 14 specific chapters will look considerably different that those shown here but will still exhibit the same critical characteristics in the key. Note that adults and larval stages of a taxon often look very different. The common names listed below are usually not the only ones discussed in each chapter.

1   Animal with four or less leg pairs and one or no pairs of antennae (legs and antennae may have been lost) 2

1' Animal with more than four pairs of legs, one or more pairs may be modified as claws, usually two pairs of antennae (not always obvious)..3

2(1) Animal with four pairs of legs (Fig. 13b).....................MITES AND SPIDERS.......Chapter 14

2' Animal with three leg pairs of legs (Figs. 13c, 13d)..........................INSECTS.......Chapter 19

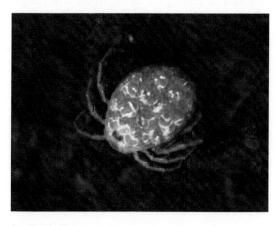

FIG. 13B. Freshwater mite (Arachnida).

FIG. 13C. Adult beetle (Insecta, Coleoptera).

FIG. 13D. Larval caddisfly (Insecta, Trichoptera).

3(1) Animal with a carapace (shell) that covers some or all of the body 4

3'   Animal without a carapace (shell), swims upsidedown, with numerous leg pairs, male with large second antennae used for clasping the female, female with a single, central brood pouch for the eggs (Fig. 13e)..FAIRY SHRIMP........Chapter 15

FIG. 13E. Egg-carrying female fairy shrimp (Crustacea, Branchiopoda).

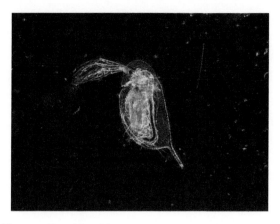

FIG. 13F. Water flea (Crustacea, Branchiopoda).

4(3)   Animal with a carapace (shell) that can close, like a clam shell 5
4'   Animal without a carapace (shell) that can open and close 7

5(4)   Carapace encloses entire animal 6
5'   Carapace covering less than entire animal, sometimes all but head and antennae (Fig. 13f).....
        ....WATER FLEAS ......Chapter 15

6(5)   Carapace flattened laterally, or if not flattened, then carapace is almost a sphere, usually dull
        in color (gray or brown) (Fig. 13g)...CLAM SHRIMPS.....Chapter 15

FIG. 13G. Clam shrimp (Crustacea, Branchiopoda).

FIG. 13H. Seed shrimps (Crustacea, Ostracoda).

6'   Carapace rounded, usually longer than broad, without growth lines, and usually brightly colored (orange, red, green, white and green, black and white) (Fig. 13h)...................................
....SEED SHRIMPS..........Chapter 16

7(4)   Body not disc shaped, lacking suckers... 8
7'   Body flat, almost circular or disc shaped, with two round suckers on the underside of the head; parasites on fish (Fig. 13i) ....FISH LICE ....Chapter 16

FIG. 13I. Fish louse (Crustacea, Maxillopoda, Branchiura).

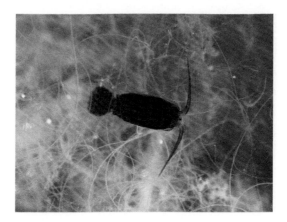

FIG. 13J. Copepod (Crustacea, Maxillopoda, Copepoda).

8(7)   Compound eyes present, or if lacking, body length more than 7 mm. If body length is less than 7 mm, then animal crawls on bottom and is never planktonic 9

8'   Compound eyes lacking, only a single central (larval) eye present; body length less than 7 mm, head and thorax larger than abdomen, antennae usually longer than head and thorax; planktonic and often red in color, or if benthic, then color is usually white or blue green, rarely red or pink (Fig. 13j)..COPEPODS.......Chapter 16

9(8)   Abdomen not flattened and folded under body 10

9'   Abdomen flattened dorsal ventrally (top to bottom), and folded tightly beneath body; body wider than long; first pairs of legs are claws (Fig. 13k) ....
....FRESHWATER CRABS..................Chapter 18

10(9)   Carapace (shell that covers the head and thorax, but not the tail) present...........................11

10'   Carapace lacking (all body segments can be seen) 13

11(10) Five pairs of legs..12

11'   Seven pairs of legs (Fig. 13l)...................... OPOSSUM SHRIMP ....Chapter 17

FIG. 13K. Fiddler crab which lives in estuarine and some freshwater habitats (Crustacea, Decapoda).

FIG. 13L. Opossum shrimps (Crustacea, Peracarida, Mysida).

FIG. 13M. *Cherax* (Crustacea, Decapoda), a freshwater crayfish from Australia which is now invasive within North America.

12(11)   Rostrum (part of shell that projects forward between the eyes and the antennae) flattened
            dorsal ventrally (top to bottom) (Fig. 13m) ....CRAYFISH.......Chapter 18
12'   Rostrum flattened laterally (side to side) usually with teeth along the top margin
            (Fig. 13n)..FRESHWATER SHRIMPS........Chapter 18

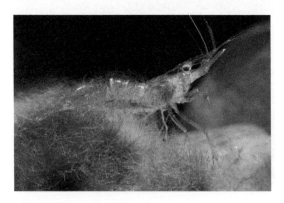

FIG. 13N. Freshwater shrimp (Crustacea, Decapoda).

FIG. 13O. Freshwater sow bug (Crustacea, Peracarida, Isopoda).

FIG. 13P. Freshwater scud, or sideswimmer (Crustacea, Peracarida, Amphipoda).

13(10) Body flattened dorsal ventrally (top to bottom); animals usually crawling, rarely
swimming (Fig. 13o)...AQUATIC SOW BUGS..........Chapter 17
13'   Body flattened laterally (side to side); animals commonly swimming as well as crawling
(Fig. 13p)...SCUDS....................................Chapter 17

Chapter 14

Mites and Spiders: Subphylum Chelicerata, Class Arachnida

I. INTRODUCTION TO THE SUBPHYLUM

Freshwater arachnids represent a small portion of Chelicerata, one of the most diverse of the living arthropod subphyla with about 100,000 described species. Their distinctive mouthparts (paired chelicerae and pedipalps) and four pairs of walking legs separate arachnids from all other arthropods. Although most arachnids and other chelicerates inhabit terrestrial ecosystems, they are an important, though largely overlooked group of freshwater invertebrates with about 5000 described species worldwide—all but a handful of which are mites. Water mites occur on all continents except Antarctica.

The actual diversity of terrestrial and aquatic mites in North America is poorly known compared to other major arthropod groups because of the paucity of taxonomic specialists in acarology, but a conservative estimate points to at least 100 species north of Mexico. Many of these species seem to have coevolved with chironomid midges and many other aquatic Diptera, and consequently exhibit comparably high levels of taxonomic and ecological diversity. Most of these mite individuals and species will belong to a group of acariform mites classified by Smith et al. (2010) as Hydrachnidiae, but several other suborders have independently adapted to freshwater habitats including suborders within Acariformes and Parasitiformes. These latter groups tend to be less diverse, abundant, and variable in form and habits.

While most aquatic ecologists group all mites as "acari" and pay little further taxonomic attention to them, the group is, in fact, quite common and diverse in most freshwater habitats and can easily be the most diverse arthropod in a benthic sample. A diligent exploration of the zoobenthos from a wetland, lake, or stream can reveal as many as 2000–5000 mites/m^2 and include 50–75 species. Most freshwater mites are ectoparasites or predators, depending on the species and life stage, and their feeding mode can vary among life history stages of the same species.

Spiders are a vastly less common group of arachnids associated with aquatic habitats. Although spiders are frequently found around stream banks and some individuals or even pseudo-colonies may spin webs across small streams, no species are considered truly aquatic. However, some members of the nursery web spiders (family Pisauridae) are often associated with aquatic habitats. In the temperate zone of North America, the only semiaquatic spider genus is *Dolomedes* (Pisauridae). Like all true spiders, members of this genus are predators.

Most of the coverage below relates to the much more common mites because information on spider biology and ecology is readily available from other field guides and specialized texts. See Smith et al. (2010) for more extensive discussion of the biology, ecology, and classification of freshwater arachnids.

ISBN 978-0-12-381426-5, DOI: 10.1016/B978-0-12-381426-5.00014-4

II. FORM AND FUNCTION

A. Anatomy and Physiology

Mites: The anatomy of water mites varies slightly between life history stages (larva, nymphal stages, and adult), but all exhibit the characteristic body plan of terrestrial mites by having a body subdivided into a mouth region (gnathosoma) and a body proper (idiosoma) representing a fusion of two other segments: the cephalothorax (head + thorax) and abdomen.

The principal internal organs are located with a body cavity (hemocoel) and bathed by fluids, which are circulated by general body movements rather than being propelled through a circulatory system. Food digestion begins externally, and only fluids enter the digestive system. Respiration generally occurs by diffusion through the outer body wall. However, large mites living in standing water supplement skin respiration with a tracheal network. Mites have light-sensitive eyespots and paired lateral eyes with lens-like capsules. Tactile sensitive setae are abundant in all species.

Spiders: The bodies of spiders, including semiaquatic species, are divided into an anterior cephalothorax (prosoma) and a posterior abdomen (opisthosoma), which is connected to the prosoma by a narrow stalk (pedicel). This body plan and the usual lack of visible segmentation make spiders easy to distinguish from other arthropods, including mites and scorpions. Spiders have four pairs of legs, all on the prosoma and each with seven segments. Also prominent on the prosoma are paired, fang-like chelicerae and a pair of pedipalps, which seem almost leg-like but which are actually another type of cephalic appendage. The pedipalps of mature males facilitate sperm transfer to females. The abdomen contains the respiratory structures (book lungs), reproductive organs, and structures for spinning silk (spinnerets).

The internal anatomy of spiders differs in some ways from that of mites. Of particular note are the poison glands present in most spiders at the upper end of the chelicerae. Spider venom has evolved to paralyze their mostly invertebrate prey (primarily insects), but the poison of some species (e.g., black widow and brown recluse spiders) is also toxic to humans. *Dolomedes* has venom capable of paralyzing small fish and frogs. Spiders liquify and partially digest their food outside the body before sucking it into their digestive tract. They have well-developed tactile and olfactory senses, and those which are highly motile (like jumping spiders) have relatively acute vision and may see into the ultraviolet. Aquatic spiders, however, sense prey primarily by waterborne vibrations. These arachnids obtain oxygen with paired "book lungs" and sometimes also tube-like tracheae. Aquatic spiders, such as the fishing spider *Dolomedes triton*, carry an air bubble underwater with them which functions as an external lung. This spider absorbs oxygen from the bubble and into the tracheae. The bubble can be partially replenished from oxygen diffusing inward from the surrounding water.

B. Reproduction and Life History

In temperate North America, water mites typically live for about a year, with most of their lives spent as nymphs and adults. In the most common life history pattern, the larval stage lasts several days at most during which the larvae engorge on host fluids and increase in size 2–5 fold. Mite nymphs feed and grow throughout the summer and then transform into adults in late summer or early fall and mate almost immediately. The females enter obligate reproductive diapause until the following spring when the eggs are laid individually or in gelatinous masses on plants, wood snags, or stones, or in rare cases they are injected by short stylets into plant or mussel tissue. Most mites produce anywhere from a dozen to hundreds of eggs, but the range is quite variable within water mites as a whole. Life history stages are complex in many mites because they are dependent on the holometabolous developmental stages (egg, larval, pupal, and adult stages) of insects with aerial adults. A few taxa of mites are multivoltine, reproducing multiple times in 1 year.

Spiders: The average life span for a spider ranges from a few weeks for many males up to a year or two for females, depending on the species and both geographic location and climate. Spiders face inherent problems in that are all predaceous, and thus females sometimes eat their potential or chosen mate. To avoid this untimely end, males use auditory, visual, and/or vibratory cues to calm the female and then inject the spermatophore through their palps into the female genital tract. Once mating has occurred, the male must depart rapidly or he will be consumed by the female. Some males are protrandric and eventually develop into females if they survive the mating experience.

III. ECOLOGY, BEHAVIOR, AND ENVIRONMENTAL BIOLOGY

A. Mite Ecology

These arachnids occur in all inland water habitats from the arctic to the subtropics of North America, including wetlands, freshwater springs, caves, ponds, lakes, small streams, large rivers, and even hot springs. Some can be found swimming or crawling among rocks and vegetation, while others live around dead organic matter or within interstitial environments. Among the more unusual habitats where aquatic or semiaquatic mites occur is the wet litter at the edges of permanent bodies of water where the mites crawl over the surface film in search of food or hosts. Moreover, species of the mite genus *Unionicola* are obligate parasites or commensals within the mantle cavity of bivalve molluscs. The habitats where larvae through adult mites are found are related to the needs of those life stages. Larval habitats and behavior tend to enhance acquisition of, and dispersal on hosts, while those of nymphs and adults promote feeding, growth, and reproduction. Dispersal among habitats on hosts has spurred development of new life history, behavioral, and morphological traits and subsequent speciation.

Larval mites are generally ectoparasites, while nymphs (starting with the deutonymphs) and adults are predators. Although few ecological evaluations have been conducted of their impact, mites almost certainly affect significantly the size and structure of many insect populations. Many aquatic ecologists tend to lump all mites into one group and thereby misinterpret or undervalue their ecological impacts in aquatic communities. It is known, however, that parasitism by water mites can reduce foraging time, behavior (e.g., territorial behavior in adult damselflies), mate success, and egg production of their insect hosts. Mites parasitize a high percentage (20–50% is common) of the insect hosts in an ecosystem, with individual adult insects infested by one to many species of mites and at total densities ranging from tens to as many as a thousand per host. The predaceous nymphs and adults are voracious predators on a diverse range of small invertebrates, such as seed shrimp (ostracods), copepods, water fleas (cladocera), and larval midges. Other species are partial scavengers. Estimates of the effects of predation are few, but some studies suggest that mites can account for half the mortality of some water flea and midge populations in ponds and small lakes. Unfortunately, their impact is rarely considered by lentic ecologists, most of whom work in the deeper, pelagic area of larger lakes.

Water mites are consumed by a variety of predators such as insects, fish, and turtles, but predation as a source of mortality seems to be less than one would expect. Adults of many species exude a sticky fluid when handled roughly, and this could deter predators. Fish tend to avoid brightly colored mites, but this aposematic coloration in mites may have originally evolved in some species as a partial response to damaging ultraviolet radiation in shallow ephemeral ponds where fish are typically absent.

B. Spider Ecology

Fishing, dock, or raft spiders, as the aquatic spiders of temperate North America are called, are large, long-legged spiders with flattened, mottled bodies. They live in association with streams, ponds, and wetlands, but they often spend considerable time away from water. Even the well-known fishing spider *D. triton* follows this habitat pattern. Fishing spiders can most readily be found on rocks, wood snags, or emergent plants at the water's edge.

Such spiders have hydrophobic hairs, which allow them to run across the water surface or dive into the water in search of prey or to escape predators without flooding their respiratory system. While they can dive quite deep, they are mostly found in the first half meter where prey are more common. They can stay submerged for half an hour using oxygen present in bubbles they carry on their ventral surface. Even though they can dive in search of invertebrates, small fish, and tadpole prey, much of their food comes from the surface where terrestrial insects become trapped at the air–water interface by surface tension. Consequently, they typically remain motionless with the anterior legs resting on the water surface in order to detect the vibratory or chemical cues, which indicate the direction, proximity, and possibly size of potential prey. By remaining motionless, these spiders also reduce risk of predation from fish, birds, frogs, and even other spiders. Small to intermediate-sized juvenile *Dolomedes* are also preyed upon by surface dwelling insects such as water striders (Gerridae) and backswimmers (Notonectidae), and predation by other spiders can be high, with cannibalism accounting for 5% of the prey of female fishing spiders. The chances of falling victim to other predators are reduced by the nocturnal habits of *Dolomedes*. This behavior also increases encounter rates with the spider's prey, which tend to be more abundant and motile at night.

IV. COLLECTION AND CULTURING

Fishing spiders can be caught at the surface in still water areas with a sweep net using a keen eye, stealthy approach, and a fast hand. They can be kept in a shallow sealed (to prevent escape) aquarium/terrarium with a combination of dry and wet habitats. Feed them various small insect prey.

Water mites, while much more abundant, are also much more tedious to catch in some ways. If you want to conduct a quantitative search, you will need a series of sieves (2–250 mm) to separate the prey from the surrounding sediment and living and dead organic matter. However, you can also just collect this material, place it in clear water within a white tray, and use a magnifying glass to spot the mites in a nonquantitative fashion. Some mites are bright red (advertizing their unpalatability to predators) and easily spotted swimming in the pan. Maintaining mites in the home or lab for long periods is difficult even for specialists because of the ectoparasitic nature of various life stages.

V. REPRESENTATIVE TAXA OF MITES AND SPIDERS: CLASS ARACHNIDA

A. Mites

Mites (Fig. 14a): Acari

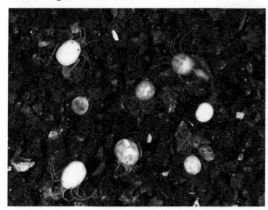

Identification: Spherical, to elongate body, without obvious segmentation. White, green, blue, brown, black, or red. **Size:** 0.5–2 mm. **Range:** Found throughout North America. **Habitat:** Fresh and brackish water. Ponds, lakes, rivers, streams. Often under rocks, or in dense vegetation. Occasionally parasitic on an insect host. **Remarks:** Cannot parasitize humans.

Red Mites (Fig. 14b): Elyias sp.

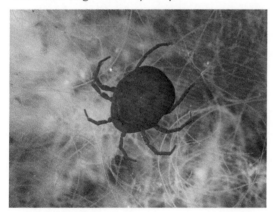

Identification: Spherical, bright crimson or crimson and white mites. **Size:** 0.8–2 mm. **Range:** Found throughout North America. **Habitat:** Freshwater, ponds, lakes, slow moving rivers, streams. Typically swimming in open water or attached to an insect host. **Remarks:** Adults are predators, larvae are parasites. Cannot parasitize humans.

Vernal Pool Mites (Fig. 14c): Undescribed species

Identification: Spherical, covered in a velvety coating, colored brown or red. Pedipalps elongated. **Size:** 0.5–0.9 mm. **Range:** Found throughout western North America. **Habitat:** Seasonal wetlands. These mites live on the surface of the wetlands and live in large groups. **Remarks:** Predatory mites that attack and feed on insects that fall into the water. The mites use their elongated pedipalps to feel vibrations or ripples made by drowning insects across the water surface.

B. Spiders

Fishing Spiders (Fig. 14d): Dolomedes sp.

Identification: Small- to medium-sized spiders with brown bodies that have patterns of white spots and stripes. Legs green to brown. **Size:** 5–25 mm. **Range:** Found throughout subarctic North America. **Habitat:** Calm freshwater, typically ponds, sheltered parts of lakes and rivers. **Remarks:** May grab aquatic prey and pull them up through the water surface.

Chapter 15

Fairy Shrimp, Tadpole Shrimp, Clam Shrimp, and Water Fleas: Subphylum Crustacea, Class Branchiopoda

I. INTRODUCTION TO THE CLASS BRANCHIOPODA

Branchiopoda is one of four crustacean classes occurring in freshwater, the others being Maxillopoda, Ostracoda, and Malacostraca (see Chapters 16–18). From a morphological perspective, the Branchiopoda appears to be a very heterogeneous group; however, they share the following characteristics: (i) the same basic larval morphology; (ii) thoracic legs (called phyllopods) that are flat, edged with setae, not distinctly segmented, and usually unbranched; and (iii) rolling mandibles that are simple and unsegmented with corrugated grinding inner surfaces. The branchiopod orders are distinguished by aspects of their trunk limbs and mouthparts. They are often separated into two arbitrary, non-taxonomic groups consisting of water fleas (order Diplostraca, suborder Cladocera) and the so-called "large" branchiopods (some of which are smaller than cladocerans) comprised of four groups: (a) fairy shrimp (order Anostraca); (b) tadpole or shield shrimps (order Notostraca) (Fig. 15a); (c) smooth clam shrimp (order Laevicaudata); and (d) spiny clam shrimp (order Diplostraca, suborders Spinicaudata and Cyclestherida).

As a group, branchiopods are present on all continents (including Antarctica), where they may be found in the open water plankton, in nearshore habitats among aquatic weeds, and on the bottom among decaying organic matter. All major groups of branchiopods occur in ephemeral pools and saline lakes lacking fish predators. In addition, water fleas also occur in lentic environments from small to large lakes where fish also reside. Species tend to differ between open water habitats and shallow, often weedy littoral zones near shore.

Branchiopoda is a relatively diverse group of invertebrates in North American inland waters. Scientists have identified about 645 water flea species in 17 genera. The highest diversity of planktonic cladocerans occurs in the formerly glaciated, mid-temperate zone, while diversity for benthic species peaks in the southeastern USA. In contrast, large branchiopods are a relatively small group of primitive crustaceans with only about 500 known species worldwide and close to 90 in temperate North America, more than half of which are fairy shrimps. Tadpole shrimp are widespread in arctic regions and west of the Mississippi River, while clam shrimp are rare north of southern Canada. Fairy shrimp occur throughout North America, and all three groups are especially diverse in the arid west.

While the survival of many large branchiopods is endangered from habitat conversion, water fleas rarely seem to be threatened by extinction, probably because of their rapid dispersal rates via resistant

ISBN 978-0-12-381426-5, DOI: 10.1016/B978-0-12-381426-5.00015-6

FIG. 15A. The tadpole shrimp *Lepidurus lemmoni*. The small red invertebrates are copepods.

eggs and their ability to colonize a greater diversity of habitats. In contrast, nearly a fourth of all large branchiopods are known from only the site where they were first identified (= type locality), and several species have been collected only once or twice.

Interestingly enough, fairy shrimp (*Artemia* in particular) are economically important, with a $20 million a year industry. Their eggs are collected in bulk from salt lakes, stored dry, and later hatched as live food for the pet and aquacultural trades. Vitamins, antibiotics, and medicines suspended in fat droplets can be fed to the shrimp for later transfer via predation to the target aquacultural fish. Furthermore, both the tadpole shrimp genus *Triops* and the fairy shrimp *Artemia franciscana* are sold as pets, the latter as "sea monkeys."

II. FORM AND FUNCTION

A. Characteristics of the Subphylum Crustacea

The subphylum Crustacea is much more diverse in body structure than the other members of the phylum Arthropoda because of their tendency to fuse body segments in different ways and to develop highly specialized appendages. They have a hard exoskeleton of chitin, which is often reinforced by calcium carbonate. Inward projections of this exoskelton provide attachments for the muscles. The body is divided in three regions, or tagmata: the head (cephalon), thorax, and abdomen, with the first two sometimes combined into a cephalothorax. At least one life stage has two pairs of antennae (insects have only one) and their mouth appendages include one pair of mandibles and two pairs of maxillae. The jointed appendages have two branches (=biramous) and may be present in all three body regions. They are variously modified among taxa for locomotion (walking, swimming, and burrowing), feeding, grooming, respiration, sensory reception, reproduction, and defense. Two extreme forms of appendages in adults are present in the subphylum: the lobed phyllopod appendages (as found among branchiopods) and the unbranched, segmented walking leg, or stenopod (typical of crayfish). Primary appendages are absent from the abdomen of all non-malacostracan crustaceans except Notostraca.

Most freshwater crustaceans use their antennae, maxillae mouthparts, or thoracic appendages to filter, grab, or otherwise collect food. Many feeding habits are common, including omnivory, predation, scavenging, herbivory, and filter feeding or scraping for algae and bacteria. Parasitism is rare, but cannibalism is common when newly molted individuals are vulnerable.

The open circulatory, excretory, and neural systems are similar among members of this subphylum, but respiratory systems are somewhat more variable. Crayfish, true shrimp, and other malacostracans have gills, whereas members of other orders primarily respire across the body surface or have secondary gills on appendages (some large branchiopods). More than 90% of the nitrogenous wastes are voided as ammonia. Crustaceans are generally sensitive to light, chemicals, temperature, touch, gravity, pressure, and sound. The eyes of adult malacostracans and branchiopods are compound and frequently mounted on a stalk, but the adult eye of ostracods, copepods, and branchiurans is a simple cluster of pigment cells (ocelli).

Freshwater crustaceans are primarily dioecious (two sexes) and reproduce sexually, but hermaphroditism and parthenogenesis occur sporadically among ostracods and branchiopods. Internal fertilization is the general rule. Females of many species protect their embryos either by retaining them in a brood pouch (e.g., branchiopods) or external sac (e.g., the ovisac of copepods) or by gluing them to appendages, such as the abdominal appendages in crayfish. Early developmental stages can take place entirely within the egg (embryo), or involve a free nauplius (as in copepods), or consist of gradual development of an independent organism.

Development of the young and adults requires periodic shedding of the older, smaller exoskeleton in a process termed molting or ecdysis. Rapid body expansion occurs immediately after the old exoskeleton is shed and before the new one hardens. The degree of expansion varies from 8–9% in some mysids to 83% in certain cladocerans. The molt process is physiologically stressing and can expose the organism to increased predation.

B. Anatomy and Physiology

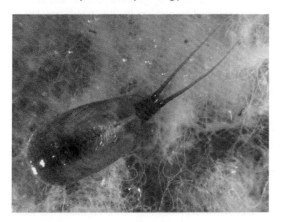

FIG. 15B. Ventral surface of the tadpole shrimp *Lepidurus cryptus.*

Large Branchiopods: This group of branchiopods is morphologically diverse and their relationships are not especially close, but they differ from cladocerans in various ways including obvious body segmentation and more pairs of legs (35–70 pairs in tadpole shrimp). While the largest cladoceran is under 2 cm in length and most are less than one-twentieth that size, fairy shrimp are usually 1–5 cm long and the predaceous *Branchinecta gigas* and *Branchinecta raptor* may reach 17–18 cm! At the opposite end, many clam shrimp are only 1.5–18 mm long. A carapace is absent in the upside-down, actively swimming fairy shrimps but present in both tadpole and clam shrimps (as well as cladocerans), which mostly crawl across the substrate and occasionally swim using their legs (tadpole shrimp) or second antennae (clam shrimp). The carapace of tadpole shrimp (Notostraca) forms a dorsal shield over the head and thorax (Fig. 15b), while the bivalved carapace of the clam shrimp enfolds the entire body, making these shrimps superficially resemble the phylogenetically unrelated seed shrimps (crustacean class Ostracoda, Chapter 39). The head of most groups are proportioned to the body size, but in smooth clam shrimp the head is massive and may be larger that the remainder of the animal. The abdomen is relatively elongated and cylindrical in both fairy and tadpole shrimp, but it is reduced to a single anal segment in spiny clam shrimp and absent in smooth clam shrimp.

Cladocera: Adult water fleas differ substantially in outward appearance from other members of this class, and many can alter their appearance from one generation to the next in response to environmental conditions—a process called cyclomorphosis. On average, however, cladocerans resemble juvenile spiny clam shrimp (in the suborder Spinicaudata) and are believed to be derived from them. Cladocerans are typically 0.2–1 mm long, but giants may reach 18 mm. Cladocerans generally have inconspicuous segmentation on the body and paired appendages (except for the second antenna). Most adults have a single central, often black compound eye (also found in copepods) and a transparent body covering (carapace), which in females is used as a brood chamber. Most also have a small, simple eye near and posterior to the large compound eye. The carapace is attached at the upper back part of the thorax and overlaps most of the body other than the head and some appendages. Water fleas cannot groom this carapace with their legs, and it can become covered with encrusting organisms in a couple of weeks if the cladoceran does not molt. The first pair of antennae is small and has a chemosensory function, while the second, large pair is used to propel the water flea. The action of these large antennae produces a jerky motion, which in combination with their shape, are responsible for their common name. The head of many species curves outward to form a beak (rostrum), and the head may be pointed and helmet-like. The shape and size of this helmet varies over generations with environmental conditions, including with the presence of some gape-limited predators. The thoracic legs of cladocerans are used as filters and collect algae electro-statically. Food retained on the legs is made into a food mass (bolus) with mucus and then moved forward toward the mouth where the entire bolus is either accepted by the mouth or expelled if it contains distasteful algae. Post-abdominal claws are used to groom these thoracic legs, and in benthic water fleas may be used to hop along the surface.

C. Reproduction and Life History

Large Branchiopods: These shrimp reproduce either sexually or by parthenogenesis, and indivi-duals may be of one sex or hermaphrodites. Sexual reproduction involves mating, which may last a few seconds to hours or days and lead to production of eggs which are stored in special brood pouches on thoracic legs. For most species, the eggs quickly develop into embryos and are released soon after development commences as resting eggs which are resistant to drought and/or freezing and viable for decades or possibly centuries without rehydration. A period of drought is needed for the eggs of some species to prevent destruction by fungus. Large branchiopods in ephemeral ponds usually have only one generation per year, whereas fairy shrimp in more permanent ponds and saline lakes produce nauplii larvae every month or so during the reproductive season. Parthenogenetic production of young also produces resting eggs.

    Cladocera: Water fleas reproduce either sexually or asexually, depending on environmental conditions. It is not surprising, therefore, that the two genders are morphologically similar, though not identical, with males being smaller than the much more abundant females. Sexual reproduction involves copulation among males and females, which can last 8–10 min—a lengthy period for such a small organism and one that makes them more vulnerable to predators. Asexual reproduction (parthenogenesis) is the norm most of the time, and the resulting "resting eggs" typically develop into females. When environmental conditions favor sexual reproduction, however, males use their post-abdomen to insert spermatozoa into the female's brood chamber. Females can produce three types of eggs. Subitaneous eggs develop immediately into neonates (sexually mature larvae) inside the mother's brood chamber and are released when the mother molts. Resting eggs can be produced sexually or asexually and are designed to enable survival under periodic or unexpected, stressful environmental conditions. These embryos are commonly extruded into a thickened carapace called the ephippium which when released is resistant to freezing or drying. Resting eggs usually develop into females, which then produce more young by parthenogenesis. The presence of predatory fish leads to increased eggs production but decreased adult sizes in the next generation.

III. ECOLOGY, BEHAVIOR, AND ENVIRONMENTAL BIOLOGY

A. Habitats and Environmental Limits

Large Branchiopods: These branchiopods inhabit many types of temporary to permanent pools and lakes that typically lack fish capable of eating adult branchiopods. In addition, these crustaceans live in temporary habitats peripheral to permanent lakes, which cannot be accessed by predaceous fish. Some of their habitats include ephemeral to semipermanent playas, vernal pools in general (from rain or snow-melt), tundra and alpine pools, rock pools (tinajas and gnammas), arctic pools that never dry but freeze solid for much of the year, pools in salt flats and alkali pans, relatively permanent alkali lakes, and various sizes of small, medium (e.g., Mono Lake in California), and large salt lakes (Great Salt Lake in Utah). These ephemeral systems are generally dry for most of the year or even in some cases for multiple years.

    Fairy shrimp are mostly planktonic but can be found throughout the water column. Clam shrimp will actively swim but spend most of their time on the bottom on in vegetation if present. Tadpole shrimp are primarily benthic and even burrow shallowly in the substrate; however, they will swim to avoid predators or beat the surface for oxygen if the water becomes depleted.

Given the habitat preference of large branchiopods, it is not surprising that physicochemical constraints are somewhat different from those of other crustaceans. Those that live in more permanent salt lakes face different constraints from most other large branchiopods. They are exposed to relatively permanent conditions over the life span of an individual but extreme salt concentrations compared to other inland habitats and even marine systems. In contrast, species inhabiting ephemeral pools and wetlands face extreme fluctuations in environmental conditions over the life of an individual and sometimes on a daily or weekly basis. Oxygen and carbon dioxide concentrations fluctuate daily with photosynthesis from resident algae or aquatic plants, and this alters pH by changing concentrations of carbonic acid in these shallow pools. Rainfall also alters concentrations rapidly because of the low volume in these ecosystems. As the ephemeral pools lose volume from seasonal evaporation, the ionic and gas concentrations change in absolute amounts and in daily variability. It is not surprising, therefore, that most large branchiopods in ephemeral habitats are osmoconformers (i.e., able to tolerate highly variable internal salt concentrations) while species in salt lakes are extremely efficient at regulating internal salt concentrations. These large branchiopods also face thermal problems because temperatures fluctuate widely in the shallow pools, especially in arid regions. Many can survive temperatures as high as 35°C, dying if the temperature drops below 25°C, and some can tolerate 40° C. Others are winter taxa, even swimming beneath ice and living at or near 0°C.

Cladocera: As indicated previously, water fleas can be found in a wide diversity of ephemeral to permanent pools, wetlands, and lakes, and they occur in medium sized streams to the largest river (especially in slow-moving lateral areas). Scientists have focused mostly on the planktonic species, but the largest diversity occurs in littoral zones and benthic habitats.

Cladocerans in open waters migrate vertically 6–8 m on average in lakes on a daily basis, moving upward after dusk and downward at dawn. The lower limit is thought to be set by the depth of light penetration and/or the amount of dissolved oxygen. There is also a detectable movement toward the littoral zone at dusk and away at dawn in some pelagic species. The majority opinion among scientists is these daily migratory patterns help the crustaceans avoid fish predation during the day, maximize access to algal food at night near the surface, and stretch energy reserves by living most of the day in cooler deep waters where metabolic rates are lower.

From the standpoint of physiological tolerance, water fleas colonize a broad range of habitats. Few Cladocera can live at temperatures above 30°C, and those native to colder regions are even more sensitive to high summer temperatures. Interestingly enough, exposure to a diet of blue-green algae (cyanobacteria) also lowers thermal tolerance. Cladocerans in general thrive best in neutral to alkaline waters (rarely acidic), in systems with low turbidity where particles do not interfere with filtration (hence, they are rare in sediment-loaded prairie rivers), and at medium levels of algal productivity (mesotrophic systems). Some species have been found in extremely soft waters (nearly distilled levels), while others survive at salinities greater than seawater.

B. Functional Roles in the Ecosystem and Biotic Interactions

Large Branchiopods: The great abundance of large branchiopods in many ephemeral pools and salt lakes makes them major components of the ecosystem in terms of food webs and nutrient cycling, especially because of the lack of fish in these systems. Densities of 200 or more animals per liter are not uncommon. Fairy shrimp, which are mostly planktivorous, are generally omnivorous filter feeders, though some species prey on rotifers and other microcrustaceans. Two giant species, *B. gigas* and *B. raptor* are exceptions, in that they feed on other fairy shrimp, clam shrimp, cladocerans, and in the latter species, on amphibian larvae. Tadpole shrimp, which are predominately benthic, are also omnivorous but focus more on detritus, scavenging, and predation than most fairy shrimp species. Clam shrimp feed on detritus and algae in suspension or scraped from plants or off the bottom.

Because of their vulnerability to predators, large branchiopods typically occur in habitats without fish, but they are still preyed upon by other species. Predators include birds (the major predator in salt lakes), amphibians, shrews, some insects, and other branchiopods. The branchiopods are sufficiently abundant so that many aquatic birds can simply filter the water to obtain massive numbers of these crustaceans. Predation also occurs on the resting eggs of branchiopods. As far as scientists presently know, neither viral, bacterial, nor protozoan parasites/diseases seem to be major problems for large branchiopods.

Cladocera: Although some water fleas are predators or specialize on eating dead organic matter (detritivores) and a very few are ectoparsites, the majority consume microscopic algae floating in the water or attached to some surface. "Herbivorous" cladocerans consume particles as small as bacteria a micron long, 1–25 µm long algae (except the often toxic or unpalatable blue-green algae which they avoid), ciliate protozoa, small rotifers, and the larval stage (nauplii) of copepods. This material is filtered from the water column or scraped from the surface of plants or the bottom. In addition to the predaceous habits of many so-called herbivorous water fleas, other benthic and planktonic cladocerans are full-time predators on microorganisms including other water fleas, small midges, and copepods. *Leptodora* is a particularly large and voracious predator, as are the invasive spiny water flea and fishhook water flea.

Scientists have demonstrated the significant importance of biological interactions in controlling density and possibly diversity of Cladocera. For example, there are many examples of one cladoceran species reducing population densities of another. Also, planktivorous fish predators tend to exclude large-bodied Cladocera, while invertebrate predators like phantom midges and backswimmers are especially detrimental to small-bodied species. Finally, rates of parasitism by microorganisms such as bacteria, microsporidia, and fungi can be high and lethal for water fleas.

The predators of water fleas are numerous and include many sunfish, large paddlefish in rivers, nymphs of dragonflies and damselflies, phantom midges, flatworms, hydra, predaceous copepods, and other predaceous cladocerans and giant fairy shrimp. Consequently, water fleas are a very crucial link in transferring energy up algae to higher consumers.

IV. COLLECTION AND CULTURING

Many water fleas can be collected from open water or shoreline areas of permanent ponds and lakes or from slowly-moving or still waters of rivers, especially in areas lateral to the main channel. Collect the cladocerans with a slightly weighted, fine mesh net attached to a rope. Throw the net out as far as possible, allow it to sink a bit, and then pull the net to the shore, avoiding the muddy bottom. Littoral zone species can be collected with a fine mesh net or they can be obtained from weedy areas or on the surface of the bottom with a bucket. Pour the contents through a series of large and small-mesh sieves to remove vegetation and then sort the contents in a water-filled white tray. Avoid getting mud in the bucket if possible. Look for large species with a hand lens or small species with a microscope from the water you have collected and sieved.

Be cautious in collecting adult branchiopods from ephemeral pools (unless you have a permit) because several federally protected species occur in California and Oregon. Non-protected species may be collected with a fine mesh net pulled through the water or swept by hand. You can even use a large-bore pipette to suck up the animals. Examine the contents of the net in a clear container with a hand lens, and then return the animals to the pool. Another approach is to collect the dirt from a dried habitat, add water, and rear the animals that hatch from the resistant eggs.

To preserve branchiopods, kill them rapidly with 95% alcohol (ethanol is best) and then place them in sealed containers with 75–95% alcohol.

Water fleas can be cultured in the home or lab with relative ease, although entire cultures have an annoying tendency to die suddenly for no apparent reason. Therefore, it is better to maintain multiple small cultures rather than one big one and even to keep the cultures in different rooms. Note: use of formalin in the same room with these cultures can eventually kill them! When bringing cladocerans from the field, try to avoid exposing them to air or bubbles in the water; because if a bubble gets under its carapace, the animal will become stranded at the air–water surface. Allow the cultures to equilibrate to inside temperatures in the water from which they were collected. They can be fed mixed algae from a green-water fish aquarium or with algae you have cultured from species obtained from a biological supply house. Feed them enough so that the water is a faintly murky green but not opaque and replenish the food as the water clears. For more information on culturing water fleas, see Dodson et al. (2010).

Rearing some large branchiopods is relatively easy. Eggs from fairy shrimp and tadpole shrimp are available commercially as dried eggs, and these can be easily cultured in the lab or at home. Fairy and clam shrimp will eat commercial tropical fish food, and tadpole shrimp can be fed small earthworms, dried cladoceran fish food, or other tadpole shrimp. Consult the various recipes for culture media and diets written for *Artemia* when raising these saltwater species.

V. REPRESENTATIVE TAXA IN THE CLASS BRANCHIOPODA

A. Brine and Fairy Shrimps: Order Anostraca

San Francisco Brine Shrimp or "sea monkey" (Fig. 15c): *Artemia franciscana*

Identification: White to pink or rarely red fairy shrimp, males with large triangular second antennae, females with a rounded brood pouch containing eggs. **Size:** Up to 8 mm. **Range:** Western and central Canada and USA, northern México. Widely introduced. **Habitat:** Permanent or temporary salt lakes and salt ponds. **Remarks:** The parthenogenic brine shrimp has been introduced to Great Salt Lake in Utah, where it occurs with the San Francisco brine shrimp. A second native species, the Mono Lake brine shrimp is endemic to Mono Lake, California.

Chirocephalid Fairy Shrimp (Fig. 15d): *Chirocephalidae*

Identification: White, green, blue, or yellow fairy shrimp often with red highlights, males typically have coiled, lamellar antennal appendages. Females have a rounded brood pouch containing eggs. **Size:** To 13 mm. **Range:** From the Arctic to mountainous regions in the Southwest, east to the Atlantic. **Habitat:** Temporary wetlands and vernal pools. **Remarks:** Members of this fairy shrimp family are widespread across North America and Eurasia.

Streptocephalid Fairy Shrimp (Fig. 15e): *Streptocephalus sp.*

Identification: White, green, blue, yellow, red, or diaphanous fairy shrimp often with red highlights. Males have a large folded "pincer-like" antennal appendage. Females have an elongated brood pouch containing eggs. **Size:** To 35 mm. **Range:** Southernmost Canada to northern Mexico, and from coast to coast. Most other species are found in the desert southwest. **Habitat:** Temporary wetlands, alkaline, turbid, playas, and vernal pools, typically active during the summer months. **Remarks:** Members of this fairy shrimp genus and genus are found mostly in Africa and North America.

Branchinectid Fairy Shrimps (Fig. 15f): *Branchinecta sp.*

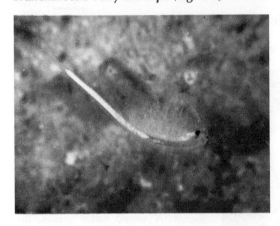

Identification: White, or white with blue, green, or yellow markings. Males lack antennal appendages and only have simple second antennae. Females have an elongated brood pouch containing eggs. **Size:** To 35 mm, except for two giant species (see below). **Range:** Found throughout western and central USA and Canada, mostly in areas with dry, hot summers. **Habitat:** Temporary wetlands, alkaline, turbid, playas, and vernal pools, typically active during the winter and spring. **Remarks:** This genus has more than 60 species in the Americas and Eurasia

Giant Fairy Shrimp (Fig. 15g): *Branchinecta gigas*

Identification: White, or white with blue markings. Large semi-predatory fairy shrimp with reduced eyes. **Size:** To 180 mm. **Range:** Deserts of western North America from southern Canada to southern California and Nevada. **Habitat:** Large temporary alkaline, highly turbid, playas and pools, active during the winter and spring. **Remarks:** the giant fairy shrimp is a predator on smaller fairy shrimp and copepods, but will also consume filamentous algae.

Raptor Fairy Shrimp (Fig. 15h): *Branchinecta raptor*

Identification: White, or white with blue markings. Large predatory fairy shrimp with reduced eyes and first leg pairs elongated for grasping prey. **Size:** To 180 mm. **Range:** Snake River Plateau, in southwestern Idaho. **Habitat:** Large temporary alkaline, highly turbid, playas, active during the winter and spring. **Remarks:** this giant fairy shrimp is a strict predator on other crustaceans, insects and amphibian larvae.

Beavertail Fairy Shrimp (Fig. 15i): Thamnocephalus sp.

Identification: White fairy shrimp often with red, yellow or blue highlights. Abdomen broadly paddle shaped. Males typically have a large branched appendage in the middle of the head. Female second antennae are extremely long. **Size:** To 55 mm. **Range:** Two species, *T. platyurus* found from Montana through the western plains states to México, west to southern California, and *T. mexicanus* from Texas, New Mexico, Arizona and northern México. **Habitat:** Temporary wetlands, alkaline, turbid, playas and vernal pools, typically active during the summer months. **Remarks:** Usually found in low numbers in wetlands.

B. Tadpole Shrimp: Order Notostraca

Triops Tadpole Shrimp (Fig. 15j): Triops sp.

Identification: Green, brown, yellow, or gray shrimp without a paddle-shaped plate between the twin tails (cercopods). **Size:** To 80 mm. **Range:** Recorded from southern central Canada south through the western plains states to Texas and México, west through Nevada and California. **Habitat:** Temporary wetlands, alkaline, turbid, playas and rice paddies. Active during the summer heat, feeding on other invertebrates and plant matter. **Remarks:** Sometimes sold in toy and pet shops in a kit for rearing them from egg.

Winter Tadpole Shrimp (Fig. 15k): Lepidurus sp.

Identification: Green, brown, black, yellow, or gray shrimp sometimes mottled, lacking a small paddle-shaped plate between the twin tails (cercopods). **Size:** To 80 mm. **Range:** Six species mostly found west of the Rocky Mountains, although one species, *L. arcticus* is found throughout the subarctic and arctic regions. **Habitat:** Temporary wetlands, alkaline, turbid, playas, vernal pools, and permafrost ponds. Active during the winter and spring, these species are opportunistic predators, sometimes cannibalizing each other, as well as feeding on plant matter. **Remarks:** One California species *L. packardi* is federally listed as an endangered species.

C. Smooth Clam Shrimp (Fig. 15l): Order Laevicaudata

Identification: Brown, sometimes slightly greenish, spherical clam shrimp, lacking growth lines. Head is large, sometimes as large as the rest of the entire body. **Size:** 2–8 mm. **Range:** Widely distributed across most of North America. **Habitat:** Well-vegetated temporary wetlands, and alkaline, turbid, playas, and vernal pools. Typically grazing on algae, or phytoplankton. **Remarks:** Closely related to tadpole shrimp.

D. Spiny Clam Shrimp and Water Fleas: Order Diplostraca

Spiny Clam Shrimp: Suborders Spinicaudata and Cyclestherida (Fig. 15m): Cyclestheria *sp.,* Limnadia *sp.,* Eulimnadia *sp.,* Cyzicus *sp.,* Eocyzicus *sp., and* Leptestheria compleximanus

Identification: Brown to yellow clam shrimp, with clear, obvious, growth lines on the carapace causing them to resemble a clam. **Size:** 5–12 mm. **Range:** Widely distributed across most of North America. **Habitat:** Well vegetated temporary wetlands, and alkaline, turbid, playas, and vernal pools. One species in permanent ponds in Florida and Texas. **Remarks:** These genera are in need of revision.

Water Fleas (Fig. 15n): Suborder Cladocera

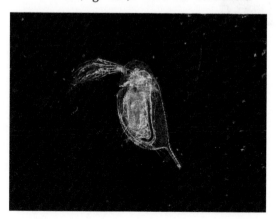

Identification: Carapace covering body. Tend to be white, yellow, blue, or red, sometimes with black markings. Growth lines are absent. **Size:** 2–8 mm. **Range:** Widespread throughout North America. **Habitat:** Lakes, ponds, playas, wetlands, temporary wetlands, and vernal pools. Filter feeds on plankton. **Remarks:** Sometimes referred to in general as "Daphnia," however, *Daphnia* is only one genus of many in this group.

Spiny and Fishhook Water Fleas (Fig. 15o): Bythotrephes sp., Cercopagis sp.

Identification: Carapace barely covering body, white. Abdomen greatly elongated, thread like, with small or large spines. **Size:** To 12 mm. **Range:** Introduced to the Great Lakes, may be spreading. Native to Eurasia. **Habitat:** Lakes, ponds. Predatory on plankton. **Remarks:** Invasive species that have changed food webs in the Great Lakes. They tangle on fishing lines by the hooks on their bodies. These animals are native to the Ponto-Caspian Region of Eurasia.

Giant Water Flea (Fig. 15p): Leptodora kindtii

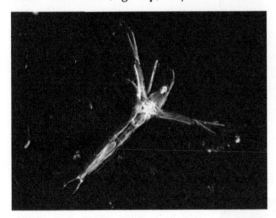

Identification: Carapace lacking, body nearly transparent. Abdomen elongated and cylindrical without spines. Its one single fused eye is situated on a long projecting head. **Size:** To 18 mm. **Range:** Widespread in North America and Eurasia. **Habitat:** Lakes, ponds. Predatory on plankton. **Remarks:** Our own native, large predatory water flea. Almost entirely diaphanous, usually is only noticed because it moves.

Chapter 16

Copepods, Fish Lice, and Seed Shrimp: Subphylum Crustacea, Classes Maxillopoda and Ostracoda

I. INTRODUCTION TO COPEPODS, FISH LICE, AND SEED SHRIMP

A. Introduction to the Class Maxillopoda

Maxillopoda is a large class of mostly small crustaceans (typically 0.5–2 mm) with over 14,000 described species consisting of freshwater and marine copepods (subclass Copepoda), marine barnacles, and a few other groups, including fish lice (subclass Branchiura). Most species are free-living, but some individuals of all subclasses and some entire subclasses are parasitic. All species have a reduced abdomen with few or no appendages. Within freshwater the class contains one very diverse group, the copepods (subclass Copepoda), and one small group, the fish lice (subclass Branchiura). We briefly discuss the fish lice below, but will spend most of this chapter discussing copepods and another Crustacean group, the diverse seed shrimps of the class Ostracoda.

Introduction to the Subclass Copepoda: Copepods are often the most abundant macro-zooplankton in lakes, streams, and oceans throughout the world and can even inhabit wet organic soils. The subclass contains about 2500 described species of free-living, mostly omnivorous copepods in freshwater habitats around the world. With exception of one free-living species of *Gelyella* (order Gelyelloida), all North American freshwater copepods are members of the orders Calanoida, Cyclopoida, or Harpacticoida. They can be found in all inland water habitats except some saline lakes. Many species are planktonic but most live near the bottom. Although copepods are typically microscopic in size (0.5–1.0 mm), some attain lengths of 2–3 mm and can be seen with the naked eye. While most of these crustaceans are transparent or a pale gray or brown, others are black or brightly colored (e.g., red, green, and blue) as a result of absorbed plant pigments held in oil droplets within the copepod.

Synopsis of the Subclass Branchiura: Somewhere around 150–200 species of fish and carp lice (subclass Branchiura) have been identified globally, with most found in tropical regions of Africa and South America and most occurring in freshwater habitats. All 22 species occurring in North America are in the genus *Argulus*, and all but three occur east of the Rockies, with the greatest concentration in the Gulf States. Tongue worms (crustacean subclass Pentastomida) are presently thought the nearest relatives of fish lice, but some systematists argue for a closer relationship with seed shrimp (class Ostracoda).

ISBN 978-0-12-381426-5, DOI: 10.1016/B978-0-12-381426-5.00016-8

All fish lice are ectoparasites, and all *Argulus* use a sheathed hollow spine and proboscis to extract blood, extracellular fluids, and/or mucus from fish. Once engorged with fluid, fish lice can skip feeding for 2–3 weeks. In most cases, this feeding mode does not cause lasting harm to the host.

Fish lice differ from copepods in having a broad and flattened body, four pairs of legs, large suckers (modified from maxillae mouth appendages), and a pair of compound eyes. Females fasten their eggs in rows to rocks or other firm benthic substrates rather carrying them in ovisacs like copepods do. Larvae hatch after 12–30 days and then swim within the lake or stream in search of a fish host. When they encounter a suitable host, they attach with cephalic hooks on to an external surface or within the gill chamber. Adults develop after 4–5 weeks and swim awkwardly (somer-saulting) through the water in search of mates and new hosts.

B. Introduction to the Class Ostracoda, the Seed Shrimps

Seed or mussel shrimp are a moderately large group of crustaceans with about 8000 freshwater and marine species described so far. Around 420 species have been identified from North America. These ostracods (also spelled ostracodes) can be distinguished from most other freshwater crustaceans by a protective, dorsally attached, bivalve shell which lacks growth lines found in the much larger clam shrimp (Chapter 15). Freshwater species rarely exceed 3 mm, while those in North America are usually 0.4 mm to a bit over 1 mm long. Ostracods are basically benthic species because of their heavy shells, but some are occasionally collected in plankton. All are free-living except some in the family Entocytheridae, which are commensal upon other crustaceans. Seed shrimps can be found in all inland water habitats from ephemeral to permanent, surface to caves, freshwater to saline, and ambient to hot spring thermal habitats.

II. FORM AND FUNCTION

A. Anatomy and Physiology

General information on anatomy and physiology of the subphylum Crustacea was discussed in Chapter 15, and this applies to the two classes covered here. Therefore, the following coverage only summarizes some key elements.

Copepods: Copepods are easily distinguished from other crustaceans. Their somewhat cylindrical body has obvious segmentation, abundant, segmented appendages on head and thorax, two spiny (setose) nonlocomotory appendages (rami) projecting from the last abdominal segment, and a single, large eye on the head. Their first of two pairs of antennae are very prominent and assist in locomotion, feeding, and reproduction. Their thoracic, swimming legs are important taxonomically and are the source of the subclass name (from the Greek *kope* and *podos*, meaning "oar foot"). The highly visible, anterior eye is a simple eye incapable of forming images, unlike those in some of the more complex crustacean relatives.

Seed Shrimps: Ostracods have a shortened body with a slight constriction between the head and thorax; an abdomen is missing. The most prominent feature of seed shrimp is their bivalve shell which is composed of chitin, heavily impregnated with calcium carbonate in most species. It is an extension of the cuticle and is molted frequently as the animal grows. The four pairs of head appendages are used for swimming, walking, and feeding, while the two pairs of thoracic appendages are employed for feeding, creeping, and shell cleaning. The internal anatomy of seed shrimp is simplified in accordance with their small size. Both heart and gills are lacking, with respiration occurring across the entire body surface. The single median eye is composed of three optic cups, each containing lenses. Seed shrimps can detect shapes and motion when the shells are open but only light intensity when closed. Subterranean species, however, are typically blind.

B. Reproduction and Life History

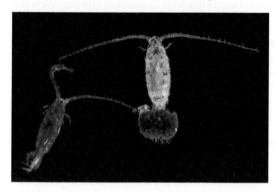

FIG. 16A. Male (left) and egg-bearing female (right) of the copepod *Aglaodiaptomus leptopus*.

Copepods: With the exception of a few species of harpacticoids, all copepod species reproduce sexually. The eggs are carried by the female or sometimes scattered in the water (Fig. 16a). Normal eggs hatch within a few days, but resting (diapausing) eggs can survive extended periods of dormancy (several years) until favorable conditions return. After the fertilized eggs hatch, the developing copepod goes through six naupliar (larval) and six copepodid stages, the last of which is the adult stage. An early stage nauplius is barely recognizable as a copepod, but successive molts gradually transform it into the adult crustacean. The entire development process following hatching lasts 1–5 weeks, and adults can typically live as long as one to several months.

Seed Shrimps: More than half the species of freshwater seed shrimps are sexually dimorphic and reproduce sexually following copulation between the generally larger males and the smaller females. Parthenogenesis, however, is quite common especially in disturbed habitats. Brood care is rare but present in the small subclass Myodocopa, whose members have a more flexible shell. A much more common behavior is to avoid brooding and simply glue the eggs to a firm surface and then abandon them. Some of these eggs are resting eggs that can tolerate desiccation and exposure to extreme temperatures. Eggs typically hatch in spring in the temperate zone. Ostracods usually pass through eight molt stages from egg to adult, at which point molting ceases. They reach sexual maturity within a few months to over a year, and may live multiple years (4 years in one species).

III. ECOLOGY, BEHAVIOR, AND ENVIRONMENTAL BIOLOGY

A. Habitats and Environmental Limits

Copepods: Copepods occur in open water, ephemeral wetlands, and nearshore areas of permanent lakes where they are important members of both zooplanktonic (mostly calanoids and some cyclopoids) and benthic communities (primarily the harpacticoids but also cyclopoids). They also occur in flowing water habitats, especially in streams and rivers and in lateral slack waters, but they are usually less abundant there than in lentic habitats. Pelagic copepods are proficient swimmers and use their first pair of antennae and sequentially thrusting legs to generate this locomotion. Benthic species tend to crawl or swim over substrates.

Copepods can thrive in ephemeral to permanent habitats in a wide range of seasonally fluctuating environmental conditions because they can reduce their metabolic rates and enter diapause in either the egg or late copepodid stages. Changes in temperature or oxygen concentration can induce diapause, as both may signal onset of unfavorable growth or survival conditions. Some copepods living in benthic sediments can survive low or even zero oxygen conditions for long periods, and a few planktonic species can tolerate these minimal oxygen conditions temporarily during their daily vertical migrations into deeper, stagnant waters. An equally or more serious threat comes from ultraviolet radiation (UVR). The intensity of UVR is a serious problem for crustaceans living in exposed, shallow water habitats, while lake populations may escape UVR by residing during the day in deep and dark waters. Intense UVR increases selective pressures to produce color morphs more tolerant of UVR, but copepods in general are relatively adept at repairing UVR-damaged DNA in their cells.

Seed Shrimps: Ostracods as a group occur in all inland water habitats on the surface and in caves wherever sufficient organic detritus exists in an undisturbed state. They are common in both ephemeral wetlands and deep, permanent lakes and rivers. Extreme habitats, such as hot springs and highly saline lakes, also are home to seed shrimp. Despite this wide distribution, however, the types of dissolved ions and their concentration (salinity) play a major role in the distribution of individual species in North America. Some ostracods frequent moist organic mats and hydrated soils of wetlands and may even colonize the axial cups of epiphytic bromeliad plants. Most species move within their primarily benthic habitat by crawling, but some species are weak swimmers.

While seed shrimps are found in a large range of salt concentrations from medium to hard water habitats, they are very sensitive to low to medium oxygen concentrations even though they can tolerate a wide range of temperatures.

B. Functional Roles in the Ecosystem and Biotic Interactions

Copepods: Copepods play important roles in aquatic food webs as primary and secondary consumers. Most are omnivorous or mostly herbivorous, consuming foods such as detritus, pollen, and phytoplankton, but some groups (especially cyclopoids) are raptorial predators on other invertebrates such as protozoa, rotifers, nematodes, and midge larvae. Some large copepods can capture small larval fish. As a very abundant prey items for larger invertebrates and fish, copepod larvae and adults are a vital food web link between phytoplankton and higher level consumers.

Copepod populations seem to be controlled by both food availability and predators. The latter include other copepods (especially cyclopoids), larvae of the phantom midge *Chaoborus*, the exotic cladoceran *Bythotrephes cederstroemi*, many fish species, and some other vertebrates.

Seed Shrimps: Most ostracods feed on algae and/or organic detritus. Carnivory and parasitism are present but very rare in this class. Bottom-feeding fish are common predators of seed shrimps. Other predators include copepods, various aquatic insects, and oligochaetes. A few ostracods appear to release a toxic chemical that can quickly dissuade some attacking predators.

IV. COLLECTION AND CULTURING

A. Collection

Copepods: These crustaceans can be collected most easily with a professional or homemade plankton net attached to a line. Throw the net out as far as possible into the water and pull it back at a steady pace. The net can also be lowered toward the bottom and rapidly pulled to the surface from a boat or dock. The type of zooplankton collected will vary with ecosystem type (river, lake, or wetland), habitat sampled (open water vs. shallow, weedy shoreline areas), and time of day (for vertically migrating species). Avoid pulling the net through the mud (which clogs the net), but you can sample near the bottom to collect certain groups like harpacticoid copepods. Wash and transfer the samples from the net into a glass container (with as little water as possible) so that the copepods can be seen with the naked eye (for large specimens) or later concentrated for viewing under a microscope. Preserve specimens in 70–75% ethyl alcohol (or rubbing alcohol if ethanol is unavailable).

Fish Lice: Fish lice are most likely to be observed by chance on their hosts when one is fishing in a lake or river. They are easily seen when present.

Seed Shrimps: These crustaceans are best collected with a benthic grab sampler or a D-frame net from shoreline and then sieved through a mesh of about 0.15 mm (smaller pores are needed to collect smaller species). Pools are the best spot to look for ostracods in streams because more organic matter accumulates there. They can be collected from underground habitats by placing a fine mesh net (63 μm) in the outflow of an aquifer stream to collect the typical very small ostracods.

B. Culturing

Copepods: These invertebrates can be cultured through multiple generations in filtered lake water or in purchased spring water, but in either case make absolutely certain to avoid glassware contaminated with most soaps, acids, formaldehyde, or other toxins. Copepods can be cultured on a variety of food types, with the greatest success coming from multiple food choices. For some groups, baker's yeast, dried fish food, and algae may work well.

Seed Shrimps: These crustaceans can be raised at least for a season in an aquarium supplied with water and some substrate from the habitat where the seed shrimps were collected. Check the pH periodically and adjust it to stay near that present initially. Several optional food sources have been used successfully to raise ostracods, including cultured algae, leaf lettuce or grated boiled egg which has been decomposing for a month, benthic algae from the natural habitat, and gastropod pellets.

V. REPRESENTATIVE TAXA OF COPEPODS, FISH LICE, AND SEED SHRIMPS: CLASSES MAXILLOPODA AND OSTRACODA

A. Copepods (Fig. 16b): Class Maxillopoda, Subclass Copepoda

Identification: Subcylindrical body, with long or short antennae. One medial eye present. Animals may be red, orange, yellow, white, green, blue, or brown. Swim in short jumps. **Size:** Small, usually 0.3–3 mm, rarely to 5 mm. **Range:** Found worldwide. **Habitat:** Lakes, ponds, wetlands, rivers, streams. In plankton or on bottom. **Remarks:** These animals typically comprise the majority of the zooplankton. Male calanoid copepod.

Copepods (Fig. 16c):

Additional Remarks: Example of a female cyclopoid copepod.

B. Fish Lice (Fig. 16d): Class Maxillopoda, Subclass Branchiura: *Argulus*

Identification: Flattened body, with very short antennae, and two large ventral suckers. Two compound eyes are present. Animals may be white, yellow, brown, or gray. Swim with the head directed upwards. Sometimes swimming in deep swoops. **Size:** Grows to 25 mm. **Range:** Subarctic, temperate, and tropical North America. **Habitat:** Lakes, ponds, rivers, streams. Most often found attached to a host fish. **Remarks:** These parasites can be found on the body or in the gills of any fish.

C. Seed or Mussel Shrimps (Fig. 16e): Class Ostracoda

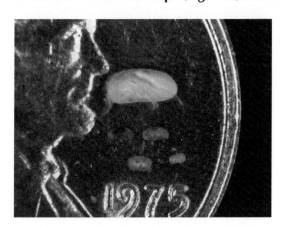

Identification: Body entirely enclosed in a shell-like carapace that lacks growth lines. This carapace may be orange, red, yellow, green, blue, white, or brown and may have stripes or spots of any of those colors. **Size:** Small, usually 0.3–4 mm. **Range:** Found worldwide. **Habitat:** Lakes, ponds, wetlands, rivers, streams. Typically on the bottom, but may swim into the lower water column. **Remarks:** Some seasonal wetland species will close their carapace tightly when the wetland dries up, and can survive until the wetland fills again.

Seed or Mussel Shrimps (Fig. 16f)

Additional Remarks: Additional examples of ostracods.

Chapter 17

Aquatic Sow Bugs, Scuds, and Opossum Shrimp: Subphylum Crustacea, Class Malacostraca, Superorder Peracarida

I. INTRODUCTION

A. Introduction to the Class Malacostraca

Malacostraca is the largest class within the arthropod subphylum Crustacea, with about 30,000 species thriving in the ocean and inland waters above and below ground. A relatively few groups have even colonized terrestrial habitats, such as coconut crabs and common pill bugs. The vast majority of crustaceans consumed by humans are in this class. At least 98% of the species in this taxon are in the order Decapoda and the superorder Peracarida. The Decapoda includes all freshwater crayfish, crabs, and true shrimp and is treated in Chapter 18. The current chapter focuses on the Peracarida, which contains about a third of the 1000+ species of malacostracans in inland waters of North America.

Common features of this class are three body regions: the head (five segments), thorax (eight segments), and abdomen (six to seven segments), each with jointed appendages, including two pairs of antennae on the head. Sometimes the head and thorax are fused into a cephalothorax, as in crayfish and crabs. The thoracic legs are variously modified for feeding, defense (with pincers), and locomotion. Abdominal appendages, often called pleopods, may be modified for swimming and sometimes respiration and care of eggs and young.

B. Introduction to the Superorder Peracarida

Six of the nine orders in Peracarida have representatives occupying permanent inland waters of North America: Amphipoda (scuds or sideswimmers; Fig. 17a), Isopoda (aquatic sow bugs), Mysida (opossum shrimps), and three orders lacking common names (Cumacea, Tanaidacea, and Thermosbaenacea). The thermosbaenaceans are very small and only live very deep underground and will not be discussed further. The Amphipoda is the most diverse peracarid order with about 300 species in subterranean and surface waters of North America. Isopoda is the next most abundant order with about 140 species, with about half inhabiting underground waters. However, the true diversity of amphipods and isopods in North America and throughout the world is undoubtedly much higher because the cave fauna is poorly known. Mysids, which are not true shrimp, are primarily marine,

ISBN 978-0-12-381426-5, DOI: 10.1016/B978-0-12-381426-5.00017-X

with only about 90 species found in surface and subterranean freshwaters around
the world, especially in the plankton of
mostly large lakes. North America is
home to 13 species of Mysida living in
fresh or brackish waters, 4 of which were
introduced from other continents.

Unlike the crustaceans groups discussed in Chapters 15 and 16, peracarids
generally lack a diapause stage and thus
have no adaptations for surviving desiccation. This characteristic and problems
moving upstream against strong currents
have limited their dispersal ability and
restricted them to mostly permanent
water bodies. Consequently, populations
tend to become genetically isolated, speciation increases, and extinction rates rise.
Moreover, groundwater pollution has
increased extinction rates for the many
species whose distributions are limited to
a few isolated cave or spring systems.

All freshwater peracarids in North
America are free-living, but external
parasitism is common in marine isopods.

FIG. 17A. Preserved specimen of the amphipod *Hyalella* sp.

II. FORM AND FUNCTION

A. Anatomy and Physiology

Peracarids share several features separating them from other crustaceans. These include a first leg
modified into a mouthpart and seven pairs of thoracic legs for grasping objects (first two pairs) or
locomotion. Also, all taxa protect their developing young in a ventral brood pouch.

The three major groups (amphipods, isopods, and mysids) are easily distinguished from each other.
Isopods appear dorso-ventrally flattened because of their seven pairs of large, thoracic appendages
(a pair of hinged-claw legs and six pairs of pointed walking legs). Their abdomen is quite broad and
shield-like, and most aquatic sow bugs are in the range of 5–20 mm long (not counting the antennae).
Amphipods also have prominent thoracic legs, but these appendages are projected ventrally, giving
amphipods a much more apparent lateral compression. Scuds are also fairly small (5–25 mm). By
contrast, mysids are generally somewhat larger (10–30 mm) than isopods and amphipods. Opossum
shrimp have small thoracic legs, a long prominent abdomen (about half the body length), and stalked
eyes (peduncle), with the last also characterizing decapod shrimp (Chapter 18).

Peracarids share many of the general features of anatomy and physiology for freshwater crustaceans, as reviewed in Chapter 15. Like other malacostracans, they breathe through gills, but these
respiratory structures are on the inner surface of the thoracic legs in amphipods and mysids but are
located at the base of the abdominal appendages in isopods. Their compound eyes are sensitive to
light intensity and motion. Their osmoregulatory abilities vary greatly depending on the habitat for
which they have evolved. They are not remarkable among crustaceans in thermal tolerance, but some
amphipods live in warm springs at temperatures regularly as high as 33–34°C, while some other
scuds are cold stenotherms and are rarely found above 15°C.

Locomotion in peracarids varies considerably among orders. Isopods are entirely benthic crawlers.
Amphipods also mostly move by crawling, but they can swim weakly using their abdominal appendages. Sometimes they turn on their sides while swimming, which accounts for their alternative
common name "sideswimmers." Amphipods may migrate upstream seasonally, often by swimming
in large groups.

B. Reproduction and Life History

Surface-dwelling peracarids typically live for a year or less (but the large amphipod *Diporeia* may live 2 or more years), while subterranean amphipods and isopods often survive to 4–6 years. The longer life span below ground reflects low mortality rates, a stable habitat, and the need to accumulate food over long periods in an organically impoverished environment.

Reproduction often peaks in the spring for surface species, but reproduction in cave species tends to be spread over much of the year in the relatively constant environmental conditions. Most species reproduce once following copulation between a male and female, but multiple broods by the same individual are not uncommon in some taxa. For example, the amphipod *Hyalella* may produce 20 broods per year. An external larval stage is absent in peracarids and, instead, the young hatch out at a late developmental stage. Female isopods carry 20–250 eggs for 3–4 weeks, while 15–50 eggs are brooded for 1–3 weeks by amphipod mothers. A female isopod living in a stream may, however, discard the brood pouch with her young if threatened by a fish predator. By sharp contrast, a female mysid broods her young for up to several months, especially in cold water.

III. ECOLOGY, BEHAVIOR, AND ENVIRONMENTAL BIOLOGY

A. Habitats and Environmental Limits

Freshwater peracarids are most commonly found in relatively cool, clean, and well-oxygenated permanent water bodies, although some species colonize thermal springs and rarely intermittent pools (a few isopods). Very few species tolerate polluted waters. Mysids inhabit some large rivers as well as deep, often cold lakes where they migrate vertically along with other zooplankton. A few species live in caves in other parts of the world, and some brackish water species are found in North America. Amphipods are found in many permanent surface water bodies, but are most abundant in streams and cave environments where they are somewhat or completely protected from predatory fish. Indeed, considerably over half the species of scuds are restricted to ground waters. This distribution reflects both the common absence of predatory fish and the fact that isolation from other water bodies tends to promote development of new species. Amphipod densities can reach thousands per square meter under ideal circumstances. Isopods are mostly restricted to cave streams and near the mouth of springs, but they also occur in other water bodies, including the shoreline areas of lakes and rivers. It is not clear why they are not more common in surface waters, but this may relate to poorer competitive ability, restrictive reproductive requirements, or susceptibility to predators. In all inland water habitats, peracarids regularly seek dark microhabitats with sufficient food resources.

Most opossum shrimps and a couple of amphipod taxa (e.g., *Monoporeia*) migrate vertically daily, and some of the latter migrate upstream seasonally. The primary reasons for vertical migration by crustaceans and other zooplankton are thought to be access to food near the surface and avoidance of visually hunting predators. However, *Hyalella montezuma*, a pelagic amphipod in Montezuma Well (Arizona), migrates upward at dusk and downward at dawn. This might be expected in a lake population, but this unique ecosystem lacks both fish predators and thermal stratification. Interesting enough, the predaceous leech *Erpobdella montezuma* follows a similar migration in search of its amphipod prey. Amphipods tend to move upstream short distances on a regular basis (= positive rheotaxis), which helps counteract the downstream displacement by currents. On a seasonal basis, however, some scud species migrate upstream in mass.

B. Functional Roles in the Ecosystem and Biotic Interactions

Amphipods and isopods are mostly omnivores which feed on benthic organic matter, but they will scavenge and also prey on other animals when possible. Some species feed heavily on plants, including watercress, which is common in the spring habitats favored by some isopods.

Mysids are more predaceous and consume zooplankton, migrating amphipods, and newly hatched fish larvae. Native subarctic opossum shrimp were introduced to over 100 northern lakes as a source of food for fish. Unfortunately, the mysids tended to avoid migrating into the upper waters where visually orienting fish occurred and instead competed with the young fish for zooplankton prey and even ate some fish larvae. In some lakes, nonnative fish flourished at the expense of native trout. *Mysis* also attacks native *Monoporeia* amphipods with which they compete for zooplankton prey. The amphipod survives in part because it can tolerate lower oxygen conditions found in deep water when the lake is thermally stratified in the summer.

Peracarids in surface waters are preyed upon by many fish, giant water bugs and other large predaceous insects, and leeches.

IV. COLLECTION AND CULTURING

Most stream and lake species of amphipods and isopods can be collected by sweeping a small mesh net (1–2 mm) through vegetation in streams or the littoral zones of ponds and lakes. They can also be obtained with box or Surber samplers, or by placing a D-net downstream and disturbing cobble substrate upstream. Traps baited with decaying organisms (such as a smashed snail) can also be used to capture them. Keep in mind that peracarids are large enough to be easy targets for fish, so search for them either where fish are rare in the water column or where aquatic weeds or other protective covers are present. Amphipods and isopods tend to be very common near the outflow of cave streams. By contrast, the fast-swimming mysids are very difficult for the amateur naturalist to collect because you need to tow a large diameter plankton net moderately fast through deep waters of large lakes and rivers.

Amphipods and isopods are relatively easily maintained in an aerated aquarium with stable temperatures as long as no fish are present. Because they are primarily omnivores and scavengers, they can exist on a variety of food types. Dead leaves from a clean pond or stream that have been conditioned in the field (i.e., in the water for at least 2 weeks) with nutritious microbial growths of bacteria and fungi are sufficient food resources for most species. This natural material can be supplemented with algal cultures or boiled lettuce or spinach. Before adding the peracarids, allow a supportive microbial community to develop in the aquarium for at least a week.

V. REPRESENTATIVE TAXA OF SOW BUGS, SCUDS, AND OPOSSUM SHRIMPS: CLASS MALACOSTRACA, SUPERORDER PERACARIDA

A. Scuds and Sideswimmers: Order Amphipoda

Scuds (Fig. 17b): Gammarus sp. and Hyallela sp.

Identification: Body compressed laterally. First antennae longer or slightly shorter than second antennae. Color dark gray to white. **Size:** 2–15 mm. **Range:** Throughout North America. **Habitat:** Lakes, ponds, rivers, streams, and estuaries. Found in marginal vegetation, under rocks, or burrowing shallowly in very loose sand. **Remarks:** These larger amphipods are important food for fish. Many undescribed *Hyallela* species.

Burrowing Scuds (Fig. 17c): Crangonyx sp., Stygobromus sp.

Identification: Body tremendously compressed laterally; extremely thin bodied. First antennae much longer than second antennae. Color dark gray to white. **Size:** 0.8–4 mm. **Range:** United States and southernmost Canada, probably elsewhere. **Habitat:** Rivers, streams, and springs. Usually in deep sand and gravels. Rarely near the substrate surface. **Remarks:** Most species live in caves or underground streams.

Planktonic Scuds (Fig. 17d): Monoporeia sp., Diporeia sp.

Identification: Body compressed laterally. First antennae much longer than second antennae. In males the first antennae may be longer than the body. Color dark gray to white. **Size:** 2–10 mm. **Range:** Northern North America. **Habitat:** Large, deep lakes. Often burrows in the bottom during the day, at night feeding in the open water. **Remarks:** These amphipods feed on plankton in the open water.

Lawn Shrimps or Land Hoppers (Fig. 17e): *Talitroides sp., Arcitalitrus sp., and Chelorchestia sp.*

Identification: Body compressed laterally. First antennae shorter than second antennae. Color dark gray to white. **Size:** 2–10 mm. **Range:** Warm coastal areas and adjacent regions. Introduced to greenhouses throughout North America. **Habitat:** Freshwater and brackish marsh. Wet or moist leaf litter or lawns, usually near streams or ponds. **Remarks:** *Chelorchestia* sp. is native to southern Florida. *Talitroides* sp. and *Arcitalitrus* sp. are native to Australia and were introduced to California with exotic plants and have spread through the plant trade.

Western Tubebuilding Scuds (Fig. 17f): *Americorophium sp., Paracorophium sp.*

Identification: Body subcylindrical. Second antennae large, used like legs. In males the second antennae may have a large spur. Color dark gray to white, with dark transverse stripes. **Size:** 2–10 mm. **Range:** Pacific Coast drainages. *Americorophium* introduced to Salton Sea and Snake River. **Habitat:** Rivers and streams with soft substrates. Often building tubes in the mud at the base of rocks or logs. **Remarks:** The amphipod *Paracorophium* is introduced to California estuaries and fresh brackish streams from the southern hemisphere. *Americorophium* is native to the west coast.

Eastern Tubebuilding Scuds (Fig. 17g): *Apocorophium sp.*

Identification: Body subcylindrical. Second antennae larger and thicker than first antennae. Color dark gray to white, with dark transverse stripes. **Size:** 2–5 mm. **Range:** Atlantic Coast drainages and the upper Mississippi River system. **Habitat:** Rivers and streams with soft substrates. **Remarks:** Recently introduced into the upper Mississippi River system, where they appear to be spreading.

B. Aquatic Pill Bugs and Sow Bugs: Order Isopoda

Aquatic Sow Bugs (Fig. 17h): Lirceus sp., Asellus sp., Caecidotea sp., Remasellus sp., Ligidium sp., and Brakenridgia sp.

Identification: Body flattened, not capable of rolling into a ball. One obvious pair of antennae. Color typically gray, but occasionally brown or white. Cercopods (tails) usually short. **Size:** 2–8 mm. **Range:** Widespread in subarctic, temperate, and tropical North America. **Habitat:** Rivers and streams, lakes, ponds, caves, and springs. **Remarks:** Numerous species, with many still undescribed.

Subterranean Sow Bugs (Fig. 17i): Calasellus sp., Columbasellus sp., Lirceolus sp., Mexastenasellus sp., and Salmasellus sp.

Identification: Body flattened, not capable of rolling into a ball. One obvious pair of antennae. Color typically white, but occasionally gray. Cercopods (tails) often very long. Eyes usually lacking. **Size:** 2–8 mm. **Range:** Widespread in subarctic and temperate North America. **Habitat:** Caves, springs, and ground waters. Sometimes in deep sands and gravels of rivers. **Remarks:** Numerous species, with many still undescribed.

Vermiform Isopods (Fig. 17j): Cyathura politula

Identification: Body cylindrical, not capable of rolling into a ball. Antennae reduced. Color typically gray, but occasionally brown or white, with barring or spots. Cercopods (tails) usually short. **Size:** 2–10 mm. **Range:** Atlantic coastal region of the United States. **Habitat:** Rivers and streams, rarely in lakes or ponds, always in lowland areas, often in fresh brackish waters or estuaries. **Remarks:** Large specimens may give a sharp bite.

Eastern Aquatic Pill Bugs (Fig. 17k): *Cassidinidea ovalis, Sphaeroma terebrans, and Antrolana lira*

Identification: Body hemispherical in cross section, capable of rolling into a ball when threatened. Antennae reduced. Color typically gray or brown. Cercopods (tails) extremely short and not generally visible. **Size:** 2–5 mm. **Range:** Atlantic coastal and Gulf Coast region of the United States. **Habitat:** Rivers and streams, rarely in lakes or ponds, or in caves. **Remarks:** One Virginia cave species, *Antrolana lira*, is protected as an endangered species.

Western Aquatic Pill Bugs (Fig. 17l): *Gnorimosphaeroma sp., Speocirolana sp., and Thermosphaeroma sp.*

Identification: Body hemispherical in cross section, capable of rolling into a ball when threatened. Antennae reduced. Color typically black, gray, brown, or mottled. Cercopods (tails) extremely short and not generally visible. **Size:** 2–10 mm. **Range:** *Gnorimosphaeroma* occurs in Pacific coastal rivers and streams. *Speocirolana* and *Thermosphaeroma* live in caves in Texas and New Mexico. **Habitat:** Rivers and streams. **Remarks:** *Speocirolana* and *Thermosphaeroma* are endangered.

Valviferan Isopods (Fig. 17m): Saduria entomon *and* Chiridotea almyra

Identification: Body flattened, not capable of rolling into a ball. One obvious pair of antennae. Color typically gray, but occasionally brown or white. Tail end elongated and maybe pointed. **Size:** 2–30 mm. **Range:** *Saduria* is found in the Arctic. *Chiridotea* occurs east of the Appalachian Mountains. **Habitat:** *Saduria* lives in arctic freshwater pools along the northern coast of Canada and Alaska. *Chiridotea* is found in deep rivers. **Remarks:** Both species also occur in marine habitats.

Parasitic Isopods (Fig. 17n): Probopyrus sp.

Identification: Females are flattened, asymmetrical animals, not capable of rolling into a ball. Males are much smaller and are attached to the female's body. Color typically brown or black with lighter markings. **Size:** 2–30 mm. **Range:** Atlantic and Gulf states, Mississippi drainage system. **Habitat:** Permanent parasites on the gills of river shrimp. **Remarks:** The microscopic larvae attack a host shrimp. If the shrimp is not already parasitized, the isopod becomes a large female. If the shrimp already has a female parasite, then the larva becomes a male and attaches to the female.

C. Opossum Shrimps: Order Mysida

Opossum Shrimps (Fig. 17o): Mysis sp., Taphromysis sp., Neomysis sp., Hemimysis sp., Deltamysis sp., *and* Alienacanthomysis sp.

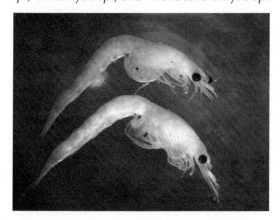

Identification: Body cylindrical, shrimp-like, with seven pairs of legs, instead of five. Female with a ventral pouch for carrying eggs. Color brown, gray, or white, occasionally red or pink. **Size:** 2–25 mm. **Range:** Generally widespread in northern North America, the Mississippi river system, and coastal areas. **Habitat:** Deep lakes; large, deep rivers, usually in brackish or fresh brackish habitats. Coastal lakes and sloughs. **Remarks:** Many invasive species from Asia occur in the Pacific coastal regions, and *Hemimysis anomola* has been introduced to the Great Lakes from the Caspian Sea region.

D. Order Tanaidacea

Tanaids (Fig. 17p): Sinelobius sp. and Leptochelia sp.

Identification: Body cylindrical, not capable of rolling into a ball. Males with very large front claws. Color typically gray, but occasionally brown or white. **Size:** 2–5 mm. **Range:** Temperate coastal California. **Habitat:** Large, deep rivers, in brackish or fresh brackish habitats. **Remarks:** The taxonomy of both genera is confused and needs revision.

E. Order Cumacea

Cumaceans (Fig. 17q): Almyracuma proximoculi

Identification: Body flattened with a long, cylindrical tail. Animal not capable of rolling into a ball. Color brown, gray, or white with markings of any of the other colors. **Size:** 2–5 mm. **Range:** Atlantic coastal and Gulf Coast states. **Habitat:** Large, deep rivers, usually in brackish or fresh brackish habitats. **Remarks:** Our only freshwater species. Numerous marine species are found worldwide.

Chapter 18

Crayfish, Crabs, and Shrimp: Subphylum Crustacea, Class Malacostraca, Order Decapoda

I. INTRODUCTION TO THE ORDER DECAPODA

The order Decapoda contains about 29,000 described species and has an estimated actual diversity ranging upwards to 110,000 species. Fewer than 500 species have been described from North American freshwaters. Over 90% of these are crayfish (Fig. 18a), while the remainder are freshwater shrimps (Fig. 18b) and crabs (most of the latter are also brackish water residents; Fig. 18c). About 80% of the world crayfish diversity occurs in North America (particularly in the southeastern USA), while in tropical countries crabs and shrimps are the most common inland water decapods. All freshwater decapods are free-living (nonparasitic).

The majority of scientific knowledge on the ecology and general biology of freshwater decapods is based on crayfish, probably because of their widespread occurrence and the fact that they are a popular food item in parts of the USA. Crabs and the freshwater shrimp *Macrobrachium* are also consumed by humans, and, in fact, *Macrobrachium rosenbergii* is perhaps the most widely cultured shrimp in the world.

The threat of extinction is a serious problem for many decapod species, especially those confined to a single or limited number of watersheds or cave environments. American crayfish have suffered from many habitat disturbances including pollution, excessive water extraction (causing desiccation of springs for example), and sometimes overharvesting. Some groups are threatened by introduction of species native to other parts of North America. An example is the Californian crayfish *Pacifastacus fortis* being threatened by *Orconectes virillis* and *Procambarus clarkii* invading from the East and *Pacifastacus leniusculus* invading from the Northwest. Around 43% of crayfish species are confined to watersheds in a single state, and 48% of crayfish species are considered imperiled to some degree. Despite these dire straits, only a handful has received federal protection under the Endangered Species Act, with a few more receiving state protection. Some shrimp are also endangered including three species of federally endangered atyid species. The Pasadena shrimp, *Syncaris pasadenae*, may now be extinct. Several palaemonid shrimp species, such as the giant Malaysian prawn *M. rosenbergii*, have escaped from aquacultural operations and are spreading in US waters where they compete with native decapods. In a counter example, native crayfish on several continents have suffered from competition and disease resulting from introduction and commercial aquaculture of the commercially harvested US crayfish *P. clarkii*. Conversely, the giant Australian crayfish *Cherax quadricarinatus* has become established in Puerto Rico, Jamaica, and northern Mexico, where it is spreading toward the southwestern USA. All freshwater crabs in temperate North America have been introduced by humans from subtropical and tropical regions of the world or have moved inland from estuarine habitats in North America.

ISBN 978-0-12-381426-5, DOI: 10.1016/B978-0-12-381426-5.00018-1

FIG. 18A. Signal crayfish (*Pacifastacus leniusculus*) native to the Columbia River drainage but widely introduced west of the Rocky Mountains.

FIG. 18B. *Palaemonetes paludosus*, a common freshwater shrimp in the Atlantic and Gulf Coast drainages.

FIG. 18C. *Rhithropanopeus harrisii*, an Atlantic estuarine crab introduced to freshwaters in Texas and coastal areas of the North American Pacific Coast.

II. FORM AND FUNCTION

A. Anatomy and Physiology

Most of the pertinent information on the anatomy and physiology of decapods was described in Chapter 15 in a section on characteristics of the subphylum Crustacea and in Chapter 17 on features of the class Malacostraca. Therefore, the following coverage is limited to select features for this order. Decapods are enclosed in a chitinous exoskeleton and divided into a cephalothorax (covered by a shield-like carapace) and abdomen. They have five pairs of thoracic appendages (called pereiopods) and six pairs of abdominal appendages (pleopods). Their walking legs are unique in being unbranched stenopods. The first one (crabs), two (shrimp), or three (crayfish) pairs of thoracic appendages bear claws for food manipulating and defense/ offense. As large crustaceans, decapods have a fairly elaborate, though open circulatory system with a dorsal heart and complex filamentous gills in the branchial chamber of the cephalothorax. Their eyes are stalked and compound.

All decapods must periodically molt their older exoskeleton in favor of a larger outer shell in a process that is physiologically and energetically demanding and which can expose the crustaceans to predators. Adult crayfishes in the family Cambaridae alternate molts between a reproductive Form I (with larger claws and altered sperm transfer gonopods in males) and the nonreproductive Form II. Females also develop wider abdomens in some species when in the Form I stage. Decapods which molt in habitats low in dissolved salts may eat their shed exoskeleton to gain sufficient calcium carbonate for the newly exposed exoskeleton.

Decapod crustaceans are among the most colorful freshwater invertebrates. This may not always be apparent because the colors of older crayfish are sometimes obscured by algae and various material, but they are readily apparent on newly molted crayfish, in which red, blue, green, or gold often appear in vivid patterns next to the duller yellows and browns of many shells. Species which have evolved underground, however, often lack pigmentation in the shell. Color differences among species reflect genetic and environmental differences. Many palaemonid shrimp seem mostly transparent, and this may reflect their need to be less visible to predators because they lack the harder exoskeleton and larger protective claws of crayfish and crabs. However at night, or in dense vegetation, these shrimp will change from being transparent to white, gray, green, or blue, or may add red, yellow, or orange spots.

B. Reproduction and Life History

Freshwater decapods live anywhere from 1 to over 60 years depending on the group and habitat. Some shrimp species have the shortest life spans of 1 (*Palaemonetes*) to 2 years (*Macrobrachium*). Surface-dwelling crayfish typically live a few years on average but some cambarids survive for 6–7 years. By contrast, estimates for the life span of highly adapted cave species range to over 60 years. The maximum age of freshwater crabs in temperate North America is presently unknown but probably extends as long as most surface-dwelling crayfish.

Freshwater decapod reproduction follows a fairly consistent pattern involving copulation, sperm storage (up to over half a year in some crayfish), fertilization when the eggs are extruded, egg attachment to abdominal appendages, development of the young on the female to an advanced stage, and release of young which resemble miniature adults. However, the mitten crab, *Eriocheir sinensis*, releases planktonic larvae that drift to estuaries to develop into small crabs that then migrate back up stream. For species living multiple years (and occasionally for those with shorter life spans), reproduction occurs more than once in an individual (iteroparity). Freshwater decapods generally produce fewer and larger eggs than their marine relatives, and this is especially true for cave species. Freshwater crabs and the large shrimp genus *Macrobrachium* migrate to estuaries to produce young (which have a lower tolerance for dilute freshwaters) and then return far upstream as adults. Adults of *Macrobrachium ohione* were once common in the Ohio River, and some freshwater crabs migrate hundreds of kilometers upstream from coastal estuaries.

Sexual dimorphism is common in decapods. Female shrimp tend to be larger than males, and only the larger individuals produce many eggs. By contrast, male crayfish (except in the west coast family

Astacidae) tend to be larger in body and claw sizes, possibly because aggressive competition among males for mates and shelter is common.

III. ECOLOGY, BEHAVIOR, AND ENVIRONMENTAL BIOLOGY

A. Habitats and Environmental Limits

With a few exceptions, freshwater decapods occur in all inland water habitats from roadside ditches and wetlands to the deepest lakes and rivers and from surface streams to caves. Some crayfish will also construct burrows on land where they can reach the underlying water table. However, decapods are absent from thermal springs, glacial outwash streams, and most saline lakes. Crayfish and shrimp are relatively rare west of the central Great Plains and are especially abundant in the southeastern USA. They are more abundant in shallow waters where refuges from predators are more common. Shrimps favor weed-choked streams, and crayfish also occur in rocky areas with little vegetation. Adult crabs and some crayfish will briefly tolerate brackish waters near the coast, but freshwaters are the preferred habitat (but see crab reproduction in the next section). The most widespread of the crayfish genera in North America are *Pacifastacus*, *Procambarus*, *Cambarus*, and *Orconectes*, with the last three being the most diverse genera in species richness.

B. Functional Roles in the Ecosystem and Biotic Interactions

Feeding by freshwater decapods as a group can be best characterized as opportunistic and omnivorous. Shrimps tend to feed more on algae and plant matter (herbivorous feeding) than other decapods. However, palaemonid shrimp tend to be predators (we have observed cannibalism in *Palaemonetes*), and scavenging is common, as shown by their attraction to traps with decaying meat. Crayfish usually prefer living or recently dead animal prey (including other young and adult crayfish) when available, but they will also readily eat plant matter. Slow moving or sessile prey, such as thin-walled snails and even zebra mussels, are preferred, but they can capture small fish (and will do so in an aquarium). Little is known about the feeding habits of freshwater crabs, but they are certainly omnivores and predators. For example, the exotic Chinese mitten crab, *E. sinensis*, heavily consumes plant detritus but has also been observed feeding extensively on larval insects. Fiddler crabs typically filter organic particles from sand and mud but will consume insect larvae.

Competition among and within species of freshwater decapods is common and well documented in crayfish. Competition for shelters and other space is especially obvious, but food competition also occurs. Males also compete for females. Shelter is especially important in protection from predators and may be a major factor in controlling population size. Shelters are vital during the molt process when decapods are very vulnerable to attack.

Fish and other large decapods are the most common predators of decapods, but other large and some small species attack them. These include water snakes, turtles, water birds, and mammals that live in (e.g., otters) or occasionally visit streams and ponds (the latter including raccoons). Crayfish try to avoid predators by seeking shelter, fleeing rapidly with back-flips of their tails, and/or taking a defensive posture with the vulnerable abdomen down and the large chelae pointed at the predator.

Decapods are also plagued by apparent commensal species (e.g., branchiobdellid annelids; Chapter 11) and by parasites.

IV. COLLECTION AND CULTURING

A. Collection

The first thing to know is that a fishing license is typically required to collect shrimp, crayfish, and crabs. Depending on the crustacean group, you can collect decapods best either by hand, with a dip or minnow net, or by using baited traps. To obtain palaemonid shrimp from clear vegetated streams, either use a funnel (minnow) trap baited with fresh uncooked meat (check daily) or sweep a long-handled net through stream vegetation during the day or at night (when a flashlight will illuminate their ruby colored eyes). In shallow water you can capture crayfish by turning over rocks that shelter them. You can also snorkel and watch for crayfish moving across the bottom, or you can place a minnow net in a creek and then disturb the upstream bed, scaring them into the net. Minnow traps baited with meat or canned pet food (cracked open to allow odors to escape) are good baits for crayfish; however, if you live in the Southeast, be careful that you do not collect unwanted snakes at the same time! Crayfish can be collected by hand, net, or trap from caves, but these habitats are better avoided unless you are working with a professional scientist or educator because many populations are easily overexploited, possibly leading to their extinction. Crayfish that burrow on land are best captured on warm humid nights using a flashlight. Search for individuals perched at the burrow mouth or foraging nearby on land. With vastly more effort and accumulated mud, you can also dig them from their burrows. The best way to collect crabs in freshwater is with baited traps, but make sure the trap entrance is large enough for the species you are seeking and not too large to invite unwanted guests.

B. Culturing

All decapods can be raised relatively easily in a laboratory or in your home as long as you avoid their escape, fouled water, and cannibalism. They will eat almost anything organic. Feed them several times per week (but not to excess) and change the water periodically if it is not well filtered. If multiple decapods are kept in the same container, then you must provide sufficient shelters (one per crayfish or crab). These can include stacked rocks, partially broken clay flower pots, PVC pipe, etc. The key is to provide a shelter large enough for an individual but small enough so that they can protect themselves from larger, aggressive intruders. Shrimp can be kept together if there is enough vegetation for shelter. Females carrying eggs or young should be isolated, and the mothers eventually separated from the periodically dispersing young so that she will not eat them. Keeping decapods with predatory fish is generally not a good idea!

V. REPRESENTATIVE TAXA OF CRAYFISH, SHRIMP, AND CRABS: ORDER DECAPODA

A. Shrimp

Dendrobranchiate Shrimp (Fig. 18d): Penaeus sp., Xiphopenaeus sp., Litopenaeus sp.

Identification: Lateral flap of first abdominal segment overlapping second abdominal segment. First three leg pairs with pinchers. Rostrum short in *Penaeus*, long in *Xiphopenaeus*. Eyes large, antennae tend to be longer than body. Color brown, gray, or white with darker markings. **Size:** 2–25 cm. **Range:** Atlantic coastal and Gulf Coast states. **Habitat:** Large, deep rivers, usually in brackish or fresh brackish habitats. **Remarks:** Many species enter freshwater or brackish water as juveniles, and move into the sea as adults.

Atyid Shrimp (Fig. 18e): Syncaris sp., Palaemonias sp., Potimirim sp.

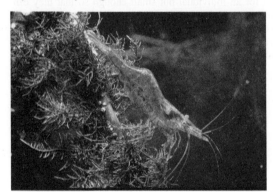

Identification: Lateral flap of second abdominal segment overlapping first abdominal segment. First two leg pairs with pinchers. First pincher pair tipped with hairs. Color black, brown, gray, mottled, or white. **Size:** 1–4 cm. **Range:** *Syncaris* is limited to California. *Palaemonias* is found in caves in the southeastern USA. *Potimirim* is native to the American tropics, and can be found in southern Florida. **Habitat:** Freshwater streams. *Potimirim* uses its claws with long apical hairs to filter flowing water. **Remarks:** *Syncaris pacifica* and the *Palaemonias* species are federally endangered, and *Syncaris pasadenae* may be extinct.

Grass or Glass Shrimp (Fig. 18b, f): Palaemonetes sp., Palaemon sp., Exopalaemon sp.

Identification: Lateral flap of second abdominal segment overlapping first abdominal segment. First two leg pairs with pinchers, and similar in size. Pinchers not tipped with hairs. Color brown, gray, mottled, or white. Often with speckle or spots of green, orange, or red. **Size:** 1–10 cm. **Range:** Widespread in the USA and extreme southern Canada. Introduced west of the continental divide. *Exopalaemon* and *Palaemon* are native to Asia. **Habitat:** Freshwater creeks and rivers, and estuaries. **Remarks:** The native North American *Palaemonetes paludosus* was introduced to the Colorado River system and the central US native *Palaemonetes kadiakensis* was introduced to California via the pet trade.

River Shrimp (Fig. 18g): Macrobrachium sp.

Identification: Lateral flap of second abdominal segment overlapping first abdominal segment. First two leg pairs with pinchers. Pinchers not tipped with hairs. Second pair of legs very robust, much larger than first pair, and often with one claw larger than the other. Color brown, gray, blue, typically mottled. Often with speckles or spots of green, orange, or red. **Size:** 5–40 cm. **Range:** Atlantic coastal states from the Carolinas south, Gulf Coast states, and the Mississippi river system. **Habitat:** Deep freshwater rivers, estuaries. **Remarks:** River shrimps were once a major commercial fishery.

Malaysian Giant Freshwater Prawn (Fig. 18h): Macrobrachium rosenbergii

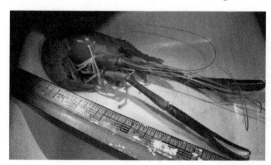

Identification: Lateral flap of second abdominal segment overlapping first abdominal segment. First two leg pairs with pinchers. Pinchers not tipped with hairs. Second pair of legs very robust, with claws extremely long. Color gray or blue (pictured specimen is preserved). **Size:** 5–40 cm. **Range:** Native to Malaysia, common in aquaculture. May escape into deep rivers in Gulf Coast states. Introduced to California, but never established. **Habitat:** Deep freshwater rivers, estuaries. **Remarks:** The most common shrimp in aquaculture today. Sometimes sold in pet shops.

B. Crayfish (Crawdads, Mudbugs, Yabbies)

Astacid Crayfish (Fig. 18a, i): *Pacifastacus sp.*

Identification: Antennae with sparse hairs or none at all. First three leg pairs with pinchers. First pair of legs very robust, with claws extremely large. Legs never with hooks on the basal segments. Color red, brown, or black. Juveniles are mottled with brown. **Size:** 5–30 cm. **Range:** Seven species and subspecies native west of the Continental Divide. *Pacifastacus leniusculus* was introduced into California. **Habitat:** Rivers, streams, creeks, cold springs. **Remarks:** The Shasta crayfish, *Pacifastacus fortis*, is a federally protected species in northern California.

Cambarid Crayfish (Fig. 18j): *Procambarus sp., Cambarus sp., Orconectes sp., Bouchardina sp., Fallicambarus sp., Distocambarus sp., Faxonella sp.,* and *Hobbseus sp.*

Identification: Antennae with sparse hairs or none at all. First three leg pairs with pinchers. First pair of legs very robust, with claws extremely large. Male's legs with hooks on the basal segments. Colors vary greatly from browns, greens, reds, and yellows with various mottling. **Size:** 5–35 cm. **Range:** Native east of the Continental Divide, with greatest diversity in the southeastern USA. Several species have been introduced west of the Continental Divide. **Habitat:** Rivers, streams, lakes, ponds, springs, creeks, ditches, bayous. **Remarks:** With over 350 species, the USA has the greatest diversity of crayfish in the world. Many species are rare or protected.

Louisiana Red Crayfish or Red Swamp Crawfish (Fig. 18k): *Procambarus clarkii*

Identification: Antennae with sparse hairs or none at all. First three leg pairs with pinchers. First pair of legs with claws extremely large, covered in tubercles. Male's legs with hooks on the basal segments. Color red, brown, or black. Juveniles are mottled with brown. **Size:** 5–30 cm. **Range:** Native to the Louisiana and adjacent areas. Widely introduced in western and parts of the eastern USA, Hawai'i, Africa, and Eurasia. **Habitat:** Rivers, streams, creeks, ponds, pools, lakes. **Remarks:** This is the most common aquaculture crayfish in the USA.

Bottlebrush Crayfish (Fig. 18l): *Barbicambarus cornutus*

Identification: Antennae with dense hairs, like a bottle brush. First three leg pairs with pinchers. First pair of legs very robust, with claws extremely large. Male's legs with hooks on the basal segments. Color mottled brown, green, and yellow. Juveniles are mottled with brown. **Size:** 5–30 cm. **Range:** The Barren River and Green River systems in Tennessee and Kentucky, USA. **Habitat:** Rivers, streams. **Remarks:** Originally treated as a species of *Cambarus*.

Cave Crayfish (Fig. 18m): *Orconectes sp., Cambarus sp.*

Identification: Antennae with sparse hairs or none at all. First three leg pairs with pinchers. First pair of legs very robust, with claws large. Male's legs with hooks on the basal segments. Eyes lacking. Live animal white to bluish. **Size:** 2–10 cm. **Range:** Subterranean streams of cave systems in the southeastern USA. **Habitat:** Subterranean streams. **Remarks:** Although the eyes may be lacking in some of these species, the eyestalks are still present.

Spider Cave Crayfish (Fig. 18n): *Troglocambarus maclanei*

Identification: Antennae with sparse hairs or none at all. First three leg pairs with pinchers. Legs and claws all very long and slender. Male's legs with hooks on the basal segments. Eyes lacking. Live animal white. **Size:** 1–5 cm. **Range:** Subterranean streams of cave systems Florida. **Habitat:** Subterranean streams. **Remarks:** This bizarre crayfish lives hanging upside down in submerged caverns. Several more species can be found in Cuba, Jamaica, and Mexico.

Dwarf Crayfish (Fig. 18o): Cambarellus sp.

Identification: Antennae with sparse hairs or none at all. First three leg pairs with pinchers. First pair of legs very robust, with claws large. Male's legs with hooks on the basal segments. Colors vary greatly from browns, reds, and yellows with black or brown mottling. **Size:** 1–5 cm. **Range:** Southeastern USA and Mississippi River system. Several species occur in northern Mexico. **Habitat:** Rivers, streams, lakes, ponds, creeks, ditches, bayous. **Remarks:** Easily mistaken for juvenile crayfish by their small size.

Parastacid Crayfish: Red Claw Crayfish, Yabby, Yabbie, Australian Crayfish (Fig. 18p): Cherax quadricarinatus

Identification: Head with four stout ridges. Antennae with sparse hairs or none at all. First three leg pairs with pinchers. First pair of legs very robust, with claws large. Male claws with a red patch on the outside edge. Male's legs without hooks. Typically blue with yellow, black, and red markings. **Size:** 5–35 cm. **Range:** Native to Australia. Introduced in northern Mexico and spreading toward Texas. Introduced in Puerto Rico and Jamaica. **Habitat:** Rivers, streams, lakes, ponds, creeks, ditches. **Remarks:** Popular in aquaculture due to its very large size. Common in the pet industry.

C. Crabs

Blue Crabs (Fig. 18q): *Callinectes sp.*

Identification: Carapace broader than long, with many marginal teeth and one large lateral spine. Last leg pair modified as paddles for swimming. Claws nearly equal in size. Typically blue or green in color. **Size:** 5–20 cm. **Range:** Atlantic and Gulf Coast lowland areas, from Nova Scotia south to Mexico. Introduced to San Francisco Bay estuary. **Habitat:** Coastal rivers and estuaries, bayous. **Remarks:** Economically important in the Atlantic and Gulf states.

Chinese Mitten Crab (Fig. 18r): *Eriocheir sinensis*

Identification: Carapace round with three marginal teeth. Claws nearly equal in size, and densely covered in hair. Usually orange or brown in life. **Size:** 5–10 cm. **Range:** Native to China. Introduced to the Sacramento/San Joaquin River system in California. Invasive in Europe as well. **Habitat:** Rivers, streams, creeks, estuaries, rice paddies, and ponds. **Remarks:** Economically important in China. This species burrows into levees, weakening them. A single Japanese mitten crab (*Eriocheir japonica*) was found on a ship in Puget Sound.

Japanese Freshwater Crab (Fig. 18s): *Geothelphusa dehaani*

Identification: Carapace round without marginal teeth. Carapace smooth. Claws greatly unequal in males. Typically orange in color, sometimes bright orange with yellow highlights. **Size:** 2–5 cm. **Range:** Native to Japan. Introduced to Lake Las Vegas in Nevada. **Habitat:** Rivers, streams, creeks, lakes, ponds. **Remarks:** Economically important in Japan. This species is a true freshwater crab, where the juveniles do not have to return to the sea or an estuary to develop. In Japan this species is an intermediate host of lung fluke. People can be infected by consuming raw or undercooked specimens.

Brackish Water Crab (Fig. 18t): *Rhithropanopeus harrisii*

Identification: Carapace wider than long, with two or three indistinct marginal teeth. Carapace with faint transverse ridges. Claws slightly unequal. Color usually brown or black. **Size:** 1–3 cm. **Range:** Native to Atlantic and Gulf Coast estuaries. Introduced to estuaries in California and Oregon and lakes in central Texas. Also introduced to Europe and the Panama Canal. **Habitat:** Estuaries, brackish to fresh brackish rivers, salty lakes. **Remarks:** This species was common in the Sacramento River Delta and Estuary until the Chinese mitten crab appeared, and has since declined, probably due to predation by the mitten crab.

Shore Crabs or Sally Lightfoot (Fig. 18u): *Sesarma sp., Armases sp.*

Identification: Carapace square in shape, with or without two or three indistinct marginal teeth. Claws slightly unequal. Usually red or reddish brown in color. **Size:** 1–3 cm. **Range:** Native to Atlantic and Gulf Coast estuaries and rivers. **Habitat:** Estuaries, mangrove swamps, brackish to fresh brackish rivers, river banks, freshwater streams, and creeks near the coast, tolerating down to 4% salinity. On land, found in grassy areas and woods near marshes, to several hundred meters inland. **Remarks:** Many species. All are excellent climbers, and may occasionally be found in trees and shrubs near water.

Fiddler Crabs (Fig. 18v): *Uca sp.*

Identification: Carapace square in shape, without marginal teeth. Eyes on narrow eyestalks. Claws extremely unequal in males. Dark brown, buff, tan, or yellowish. **Size:** 1–3 cm. **Range:** Native to Atlantic, Gulf Coast, and southern California estuaries and rivers. One species in inland lakes and rivers in Texas. **Habitat:** Estuaries, brackish to fresh brackish rivers, freshwater streams, and creeks near the coast. Salty lakes. **Remarks:** Males wave their claws to define their territories and attract mates. Primary feeding mode is to filter sand and mud for organic particles, leaving little sand pellets near their burrows.

Chapter 19

Introduction to Insects and their Near Relatives: Subphylum Hexapoda

I. INTRODUCTION TO THE SUBPHYLUM HEXAPODA

Hexapoda is the largest subphylum within the phylum Arthropoda in terms of *described* species, if not actual numbers (some arachnologists contend that the free-living and parasitic mites are more diverse, but no one knows for certain). Roughly one million hexapod species have been described from terrestrial and aquatic habitats. Of these, there are more than 10,000 species in the inland waters of North America. The subphylum name refers to the presence of six feet (three pairs of thoracic legs) present in some life stage of all hexapods. Two other characteristic are three major body regions (head, thorax, and abdomen) and a single pair of antennae. Within this subphylum are the highly diverse winged insects (class Insecta, subclass Dicondylia) and many fewer species of wingless insect (subclass Archaeognatha) and non-insect taxa (subclasses Diplura and Ellipura in the class Entognatha). The last group includes the 10–15 species of semi-aquatic springtails in North American waters; these are only briefly discussed in this book because they are only found on the surface film of quiet streams and pods. They are also smaller than most other taxa covered in this field guide.

Ten insect orders contain some aquatic species or are entirely aquatic in at least one life stage. In five orders, almost all species have aquatic larvae (called nymphs or naiads in some taxa): Ephemeroptera (mayflies, Chapter 19), Odonata (dragonflies and damselflies, Chapter 21), Plecoptera (stoneflies, Chapter 22), Megaloptera (hellgrammites and dobsonflies, Chapter 24), and Trichoptera (caddisflies, Chapter 25). The remaining five orders consist mostly of terrestrial species, but the first three can be very important components of freshwater communities: Heteroptera (Hemiptera; true bugs, Chapter 23), Coleoptera (beetles, Chapter 26), Diptera (midges, mosquitoes, blackflies, etc., Chapter 27), Lepidoptera (aquatic moths), and Neuroptera (spongillaflies). A bit over half of all aquatic insect species are in the order Diptera. Some wasps in the order Hymenoptera are technically aquatic because they inject their parasitoid eggs in the eggs or bodies of aquatic insects, but this group is not considered further in this field guide.

Insects can be found in every inland water habitat on earth where any invertebrate exists, but the more rigorous environments contain a smaller diversity and density of species. Insects have successfully colonized subterranean caves and groundwater habitats, but their numbers are more limited because the many species requiring a terrestrial life stage are limited by access to either breeding sites or food. In ephemeral to permanent pools, lakes, creeks, and rivers, insects are often the most abundant macroinvertebrate in the benthic community. A few species, such as phantom midges (Diptera, *Chaoborus*), exploit planktonic habitats throughout their larval lives, while other species are only temporarily planktonic (e.g., early life stages of larval midges) or move through the water

ISBN 978-0-12-381426-5, DOI: 10.1016/B978-0-12-381426-5.00019-3

column only in the weedy littoral zone (e.g., beetles and bugs) or live on the water surface (e.g., water scorpions). As a general rule, aquatic insects spend more of their lives in water than on land. Some species spend their entire life cycle in water, others briefly live as terrestrial adults (e.g., mayflies), and a few may live for a month or more as flying adults (e.g., dragonflies).

In these environments they represent all consumer trophic levels from herbivore to predator to detritivore. They are also extremely important to aquatic foods because of this energy transfer and the fact that they are a source of food for many predators in both the aquatic stage (e.g., sunfish and trout) and the terrestrial life stage (e.g., birds).

Aquatic insects have a mostly benign or positive interaction with humans, and as a group they rarely pose problems. However, some strong exceptions to this rule are found in the order Diptera, which includes the very annoying and often disease-transmitting mosquitoes (family Culicidae), blackflies (Simuliidae), and biting midges (Ceratopogonidae). Most aquatic stages are too small to bite/pierce a person, but the exceptions include some of the true bugs, predaceous diving beetles, and hellgrammites.

To learn more details about aquatic insect diversity and ecology, read DeWalt et al. (2010) and Hershey et al. (2010), respectively. For more detailed classification of various orders (mostly to genus), see Merritt et al. (2008).

II. FORM AND FUNCTION

A. Anatomy and Physiology

The three main body regions (head, thorax, and abdomen) are obvious in the bodies of all adults and most larval/nymphal stages, though considerably less obvious in larval dipteran flies. The head contains a number of important external structures including a single pair of antennae (obvious in most but not all larvae), a pair of compound eyes, several simple ocelli eyes, and complex mouthparts. The thorax contains three large segments (pro-, meso-, and meta-thorax from front to back, respectively), each with a pair of five-segment legs (if legs are present in the larvae). Mayflies, damselflies, stoneflies, and true bugs also feature developing wings on the thorax called wingpads. Adults in most orders have modern folded wings, but those of Ephemeroptera and Odonata are the ancient nonfolding variety. The abdomen also contains numerous segments and may have lateral segmented filaments, gills (on the thorax and sometimes head of other insects), and/or terminal appendages (cerci).

Insects respire with internal tracheae in adults, but aquatic larvae and adults may use external tracheate gills, an air bubble held over internal tracheae, integumentary respiration, or other means to acquire oxygen.

B. Reproduction and Life History

Insects usually have direct fertilization involving male–female copulation and distinct developmental stages separated by periodic molts for growth and development. Most live less than a year, with some species reproducing as often as every couple of weeks under ideal conditions, while others may live for 2 or rarely more years. Molting typically ceases when reproductive maturity is reached.

The life cycles of aquatic insects may be entirely aquatic or include a common terrestrial phase with (as in dragonflies) or without feeding adults (as in mayflies). Four insect orders with some aquatic species feature either hemimetabolous (Ephemeroptera, Odonata, and Plecoptera) or paurometabolous (Hempitera) life cycles with three developmental stages: egg, larva (or nymph), and adult. Those orders with holometabolous life cycles (Coleoptera, Diptera, Lepidoptera, Neuroptera, Megaloptera, and Trichoptera) feature four developmental stages: egg, larva, pupa, and adult. The paurometabolous nymphs of true bugs live in the same habitat as adults but lack fully developed wings and genital structures. By contrast, the nymphs of hemimetabolous orders live submerged in water, while the adults generally live on the water surface or on land and look dramatically different from the nymphs.

III.   ECOLOGY, BEHAVIOR, AND ENVIRONMENTAL BIOLOGY

A.   Habitats and Environmental Limits

Summarizing the ecology of aquatic insects or looking for general patterns is nearly impossible. Indeed, hundreds of thousands of journal articles have been written on this subject and numerous books. As indicated in the introduction, however, there are a few general but nonexclusive patterns that can be observed. From a geographic perspective, aquatic insects live in all continents (though with lower diversity and densities at the poles) from high mountain lakes to salt marshes along the coasts. All surface waters contain abundant insects, though salt lakes and thermal pools and streams have limited species (mostly dipterans). Some species colonize underground habitats (hyporheic groundwaters and subterranean caves) as long as they can successfully breed as adults, but these fauna are much more depauperate in species and numbers of individuals. Some groups, such as stoneflies, are principally found in cool, highly oxygenated waters, but other stoneflies occur in warmer, more oxygen-depleted waters of prairie rivers. Insects can occasionally survive freezing in some life stage, but can be killed by ice scour. Some groups are associated mostly with clean water (e.g., mayflies, stoneflies, and caddisflies), whereas others like many midges can thrive in organically polluted systems with very low oxygen concentrations.

Within lakes and streams, insects are most common in or on the bottom, though some species are planktonic or at least move through the water column among weeds. More individuals and species are found in the littoral zone in general than in deep water, and the fauna is remarkably different. Soft and hard substrates support dramatically different groups of species. Hard and stable soft substrates usually have greater densities and diversities than mobile substrates. And, where the latter has moderately large insect populations (in some prairie, sand-bed rivers), they are typically composed of very small individuals, with dipterans predominating. Surface-dwelling insects are confined to littoral areas of ponds and lakes or to pooled or other slow-flowing areas of creeks through rivers.

B.   Functional Roles in the Ecosystem and Biotic Interactions

The role of insects in food webs of aquatic systems is also quite diverse. They fill the functional feeding roles of detritivores, herbivores, invertivores (first-order carnivores), higher predators, and omnivores (see Chapter 4 for more detailed discussion of functional feeding groups). Their feeding modes including scraping of algae (grazers), leaf or wood shredding, plant piercing, collecting dead organic matter from the bottom or water column (=filtering or suspension feeding), and predation (whole animal biting or engulfment or consumption of liquids through piercing). These are described in more detail for individual orders in Chapters 20–27.

The feeding mode can vary substantially with habitat. For example, leaf shredders are more common in areas where the riparian canopy is extensive compared to the stream width; this is present more commonly in forested headwaters but can also occur in side channels and backwaters of large rivers. Shredders are more common in deciduous forest streams in North America than in other ecoregions, and they are even less abundant in tropical streams where crustaceans more frequently occupy this role. Algal grazers are most common wherever enough light reaches the stream bottom and the substrate favors attached algae, such as in nonforested headwaters or mid-order, rocky streams.

The biggest enemy of aquatic insects is another insect, but they are also consumed by a wide variety of insectivorous and omnivorous vertebrates, especially fish but also animals like salamanders, soft-shell turtles, some snakes, and even birds (feeding mostly on surface-skating bugs).

IV.   COLLECTION AND CULTURING

With the exception of planktonic forms like phantom midges which can be obtained at night from ponds (especially those without many planktivorous fishes) with a plankton net, many insects can be collected from streams, rivers, pools, lakes, and wetlands with either a benthic grab sampler or D-net, or by picking up rocks or other hard substrates. If the bottom contains abundant organic matter, you will need to rinse the sample through a sieve and then place the collected material on a white tray with a small amount of water. Look for all size classes of insects.

The ease of culturing insects in the home or laboratory varies with their feeding habits, life history patterns, substrate needs, and both oxygen and water current requirements. Species residing in ponds are generally easier to raise in the laboratory than those from streams because of the frequent need of the latter group for water currents to provide food and oxygen. Those with aquatic adults are often easiest to raise through a whole life cycle, but many species are suitable for study in the laboratory because their larval stages last much of the year or even multiple years, especially those from cooler climates. Keeping insects in an aquarium for extended periods is a matter of temperature control, providing oxygen and the appropriate food, and excluding predatory fish. A few groups with flying adults, such as true midges (e.g., *Chironomus tentans*, family Chironomidae) can be raised through multiple generations in gallon jars with suitable food, substrate, and mesh covers. The net covers should allow the adults to emerge and hang on to the mesh while also preventing them from escaping. The larvae can then be used to feed other insects, such as dragonfly nymphs.

V. TAXONOMIC KEY

The following key is designed to assist the user in determining which chapter to use to further identify their specimen. It is possible to reach the same taxonomic group along different pathways in the key, because of the various life stages that insects possess.

1	Obvious head present with obvious eyes	2
1'	Head and eyes not obvious	24
2(1)	Wings or wing pads containing developing wings present; wings may be hidden under a hard shell	3
2'	Wings and wing pads absent	16
3(2)	Legs present	4
3'	Legs absent.................. larval flies, TRUE FLIES: ORDER DIPTERA	Chapter 27
4(3)	Wings or wing pads not under a shell	5
4'	Wings under a hard shell .... BEETLES: ORDER COLEOPTERA	Chapter 26
5(4)	Wings developed, obvious .... adults insects	6
5'	Wing pads containing developing wings .... nymphs and pupae	7
6(5)	Mouth parts in the form of a sucking tube or beak; front half of first wing pair leathery, back half membranous .... TRUE BUGS: ORDER HEMIPTERA	Chapter 23
6'	Mouth parts not modified as a tube or beak, jaws present; first pair of wings completely leathery or hard .... BEETLES: ORDER COLEOPTERA	Chapter 26
7(5)	Legs fused to the body and not movable, although body may bend around; specimen is often in a case or cocoon or burrow, pupae	8
7'	Legs movable, nymphs	13
8(7)	Two pairs of wing pads	9
	One pair of wing pads................... TRUE FLIES: ORDER DIPTERA	Chapter 27

9(8)   Pupae in burrows or in silken, double walled cocoons, always above water line 10

           Pupae in cases made of rock, sticks or a single wall of silk, always below the water 12

10(9)   Both pairs of wing pads similar..11

10'   First pair of wing pads thickened. ..... BEETLES: ORDER COLEOPTERA ......Chapter 26

11(10)   Pupae in double walled cocoon, with outer wall like a mesh fence; small in size, less than 5 mm long.................SPONGILLAFLIES: ORDER NEUROPTERA.................Chapter 24

11'   Pupae in burrows, without a cocoon; large animals (more than 10 mm long)......................

           HELGRAMMITES AND DOBSONFLIES: ORDER MEGALOPTERA...........Chapter 24

12(9)   Jaws crossing................. CADDISFLIES: ORDER TRICHOPTERA ....Chapter 25

12'   Jaws not crossing .... AQUATIC MOTHS: ORDER LEPIDOPTERA...........Chapter 24

13(7)   Mouth parts not modified as a tube or beak...14

13'   Mouth parts in the form of a sucking tube or beak..

           .... TRUE BUGS: ORDER HEMIPTERA ....Chapter 23

14(13)   Mouth parts as jaws, not extendable..15

14'   Mouth parts folded back under head, capable of being projected out well in front of head farther out than head is long ....

           .... DAMSELFLIES AND DRAGONFLIES: ORDER ODONATA ....Chapter 21

15(14)   Gills (unless broken off) on sides of abdomen and plate or feather shaped; abdomen with three (rarely two) tails; legs ending in one claw (use hand lens)...

           ....MAYFLIES: ORDER EPHEMEROPTERA...........................Chapter 20

15'   Gills (if present) under legs or at end of abdomen and finger or bush shaped; abdomen with two tails; legs ending in two claws (use hand lens)..

           .... STONEFLIES: ORDER PLECOPTERA ....Chapter 22

16(2)   Mouth parts in the form of a sucking tube or beak 17

16'   Mouth parts not modified as a tube or beak..18

17(16)   Free swimming or living on or at the water surface; middle and often hind legs much longer than front legs; body hard . TRUE BUGS: ORDER HEMIPTERA ....Chapter 23

17'   Living on freshwater sponges; legs all about the same length; body soft, larvae.................

           .... ORDER NEUROPTERA: SPONGILLAFLIES ....Chapter 24

18(16)   Body without a hard shell, but may be in a case made of sand, gravel, sticks, leaves or silk..19

18'   Body with a hard shell not made of sand, or plant matter...

           .... BEETLES: ORDER COLEOPTERA ....Chapter 26

19(18)   Animal living under water 20

19'   Animal living on surface, usually in large groups; unable to swim, but can jump short distances ....SPRINGTAILS: ORDER COLLEMBOLA ....Chapter 19

20(19) Three pairs of legs..22
20'   More than three pairs of legs 21

21(20) First three leg pairs made up of separate segments and pointed at the tip, remaining legs
        fleshy, with a ring of small hooks (ring may be visible, but not hooks)................................
        ....AQUATIC MOTHS: ORDER LEPIDOPTERA........................Chapter 24
21'   All leg pairs fleshy, round on ends, without rings..
        .... TRUE FLIES: ORDER DIPTERA ....Chapter 27

22(20) Specimen not living in a case or a net..23
22'   Specimen living in a case made of made of silk, sand, sticks, or leaf fragments, case may
        be attached to a rock, or inside the end of a short twig, or may be living in a silk net in
        fast flowing water, using the net to capture food ....
        .... CADDISFLIES: ORDER TRICHOPTERA ....Chapter 25

23(22) End of abdomen with a pair of large hooks, abdomen with long fleshy projections,
        sometimes with a single long projection at the end of the body, jaws large..........................
        HELLGRAMMITES AND DOBSONFLIES: ORDER MEGALOPTERA ....Chapter 24
23'   End of abdomen lacking a pair of hooks, abdomen only very rarely with long fleshy
        projections, never with a single long projection at the end of the body ....
        .... BEETLES: ORDER COLEOPTERA ....Chapter 26

24(1)   Three pairs of legs................... BEETLES: ORDER COLEOPTERA ....Chapter 26
24'   More or less than three pairs of legs...
        .... TRUE FLIES: ORDER DIPTERA ....Chapter 27

VI. REPRESENTATIVE TAXA

A. Springtails (Fig. 19a): Collembola

Identification: Black, brown, gray, red, yellow, or blue. **Size:** 0.1–1 mm. **Range:** Common throughout North America. **Habitat:** On the surface of pools, puddles, and quiet ponds. Sometimes along shorelines. **Remarks:** Sometimes occur in aggregations of millions of individuals.

B. Mayflies (Fig. 19b): Insect Order Ephemeroptera

Identification: Black, brown, gray, red, yellow, or white. Maybe mottled or striped. May be flattened or elongated. **Size:** 1–15 mm. **Range:** Common throughout North America. **Habitat:** All aquatic habitats, except in caves. **Remarks:** Important food for fish and other invertebrates. Also important indicators of water quality.

C. Dragonflies and Damselflies (Fig. 19c): Order Odonata

Identification: Dragonflies lack gills, damselflies have three gills at the tail. **Size:** 1–50 mm. **Range:** Common throughout North America. **Habitat:** All aquatic habitats, except in caves. **Remarks:** Predatory insects capable of capturing other insects and even small fish and tadpoles.

D. Stoneflies (Fig. 19d): Order Plecoptera

Identification: Typically with two tails. Come in a variety of colors and patterns. **Size:** 1–65 mm. **Range:** Common throughout North America. **Habitat:** All aquatic habitats, except in caves. **Remarks:** Important food for fish and other invertebrates. Also important indicators of water quality.

E. True Bugs (Fig. 19e): Order Hemiptera

Identification: Always with piercing mouthparts, not unlike a straw. **Size:** 1–120 mm. **Range:** Common throughout North America. **Habitat:** All aquatic habitats, except in caves. **Remarks:** Includes water striders, toebiters, water treaders, water measurers, backswimmers, and boatman.

F. Hellgrammites and Dobsonflies (Fig. 19f): Order Megaloptera

Identification: Large insects with long lateral gills and large jaws. **Size:** 1–120 mm. **Range:** Common throughout North America. **Habitat:** Rivers, streams, creeks. Rarely ponds. **Remarks:** Large insect predators. The jaws look impressive, but the bite is not usually painful.

G. Spongillaflies (Fig. 19g): Order Neuroptera

Identification: White, soft-bodied insects living on sponges. **Size:** 1–5 mm. **Range:** Rare in subarctic North America. **Habitat:** Lakes, rivers, streams. **Remarks:** Mouth parts are designed to suck the contents of individual sponge cells.

H. Aquatic Moths (Fig. 19h): Order Lepidoptera

Identification: Flattened animals in cases like caddisflies. Case may be attached to a large rock. **Size:** 1–10 mm. **Range:** Uncommon throughout subarctic North America. **Habitat:** Rivers and streams. **Remarks:** This is the larvae of a true moth.

I. Caddisflies (Fig. 19i): Order Trichoptera

Identification: Worm-like insects, typically in a case made of sand or plant parts held together by silk. **Size:** 1–12 mm. **Range:** Common throughout North America. **Habitat:** All aquatic habitats, except in caves. **Remarks:** Important indicators of water quality.

J. Beetles (Fig. 19j): Order Coleoptera

Identification: Adults with hard wing covers, larvae worm-like insects. **Size:** 1–60 mm. **Range:** Common throughout North America. **Habitat:** All aquatic habitats. **Remarks:** Beetles are the most diverse group of insects in the world.

K. Midges, Mosquitoes, Blackflies, and Other True Flies (Fig. 19k): Order Diptera

Identification: Worm-like insects, typically without obvious legs. **Size:** 1–70 mm. **Range:** Common throughout North America. **Habitat:** All aquatic habitats. **Remarks:** Blackfly larvae sometimes occur in the millions in riffles and waterfalls.

Chapter 20

Mayflies: Insect Order Ephemeroptera

I. INTRODUCTION TO THE ORDER EPHEMEROPTERA

The ancient order Ephemeroptera contains about 630 described species in North America, all of which have aquatic larvae (nymphs) and terrestrial adults. The latter stage, as implied by the order name, is very short, lasting as little as a few hours to no longer than 2 weeks. Mayflies are small insects, with body sizes of the nymph ranging from 2 mm to a bit over 30 mm (exclusive of the tail and antennae). They are especially common in streams and rivers but also live in wetlands through large lakes. In all these ecosystems, mayflies are primarily either herbivores or detritivores and are prey to many species of predaceous invertebrates and fish.

Mayflies are not serious pests for humans—more like a rare and short-lived nuisance for people living by some large rivers and the Great Lakes. The periodic problem stems from massive, nearly simultaneous emergence of thousands to millions of adults of a few species of burrowing mayflies over a short period. Numbers can be so high that they make bridge roads slick and have even shorted-out insulators at electrical substations along the water body. Some people have allergic reactions to mayfly hairs and skin fragments. On the upside, however, mayflies are really important to the aquatic food web, as any trout angler can tell you.

II. FORM AND FUNCTION

A. Anatomy and Physiology

Mayfly nymphs can be distinguished from other types of aquatic insects by a combination of three characteristics: (i) the presence of three (rarely two) tail filaments, with the median filament sometimes much shorter than the lateral "cerci"; (ii) external gills attached to the side or bottom-side on most of the first seven segments; and (iii) a single claw on the last and undivided segment (tarsus) of each leg. The abdomen contains 10 segments in all species, though the first few may be concealed by an overlying dorsal portion of the thorax. Mayflies are sometimes confused with stoneflies (Chapter 22), especially when one of the tail filaments has been broken off, but stoneflies always have two claws on each leg and two tail filaments and (with one exception) lack gills on the middle abdominal segments. Species of mayflies living in fast-moving currents tend to be streamlined (e.g., *Baetis*) or dorsoventrally compressed (e.g., heptageniids), and most also seek refuge under rocks or snags to avoid unintentional downstream drift.

Adult mayflies reside during their short lives near the aquatic body that supported their nymphal stage and can be recognized by the normal presence of two pairs of clear to patterned, vertically held wings. The front wings are larger and tend be somewhat triangular, whereas the much smaller hind

ISBN 978-0-12-381426-5, DOI: 10.1016/B978-0-12-381426-5.00020-X

wings are oblong and more rounded. The hind wings are absent in some species, in which case the fore wings may be larger than normal. Two or three tail filaments extend from the abdomen; these may be longer than the remainder of the adult body.

B. Reproduction and Life History

As hemimetabolous insects, the mayfly life cycle includes three developmental stages: egg, nymph, and adult (or imago), with the nymph being aquatic and not resembling the adult. Actually, a fourth stage exists, the subimago (trout anglers sometimes call this stage a "dun"); this stage is unique to Ephemeroptera. The subimago has very fine hairs on its wings (microtrichia) which may repel water, thereby facilitating the mayfly's escape from the binding tension of the water surface. This stage may last a few hours to days before the mayfly molts to the reproductively capable adult stage (emerger or spinner to an angler). Nymphs transform into a subimago either by emerging from the nymphal skin at the water surfacing (using the old skin as a raft) or by crawling a few centimeters out of the water before transforming. Adults do not feed, with the primary function of their short adult phase being reproduction. This typically involves a mating swarm of males which attracts the fertile female. However, a secondary function relates to dispersal of the young. Female mayflies lay their eggs at an upstream site (for lotic species) on the water surface or on the underside of stones or wood snags in some species.

During the life cycle, a mayfly nymph may molt on average about a dozen times, but some species molt as many as 40–45 times. The molt pattern depends on the species, life cycle duration, ultimate body size, and prevailing environmental conditions.

The life span of mayflies varies considerably among species and climatic conditions. Some live a year and reproduce once (univoltine), while others produce multiple generations per year (multivoltine). Relatively few species require up to 2 years to develop; these are more typical of species in colder waters of springs and those found in higher latitudes and elevations.

III. ECOLOGY, BEHAVIOR, AND ENVIRONMENTAL BIOLOGY

A. Habitats and Environmental Limits

Streams are the center of mayfly abundance and diversity (especially small rocky streams with fewer large fish), but some species also occur in temporary ponds through the littoral areas of large lakes. In fact, the range of habitats occupied by mayflies is greater than the average for other insect orders. Populations of Ephemeroptera are uncommon, however, in deepwater zones of lakes. Like most invertebrates, they avoid light (and observation by predators) and thus are often found on the undersides of rocks or woody debris or within leaf packs. Mayflies primarily move by crawling over the substrate, but some species swim quite well by rapidly moving their abdomen up and down (by contrast, stoneflies swim by moving the abdomen sideways).

Mayflies vary considerably by taxa in oxygen requirements, and therefore are useful as indicators of organic pollution in stream monitoring studies.

To disperse among habitats, nymphs will frequently drift downstream, usually at night to avoid predators. Members of the genus *Baetis* are among the most common insect drifters. They can also disperse in the adult stage when the imago flies upstream to lay her eggs. Species of *Baetis* in tundra rivers of Alaska have been shown to fly 2 km upstream to deposit eggs.

B. Functional Roles in the Ecosystem and Biotic Interactions

Almost all nymphs are herbivores or detritivores, eating respectively either attached algae or dead organic matter that is well colonized with bacteria and fungi. The intermediate subimagos and terminal adult stage do not feed. Burrowing mayflies, including the relatively large genus *Hexagenia*, burrow into mud banks of streams and rivers where they undulate their bodies to pull in food and oxygen and disperse wastes. A few species of mayflies are predaceous.

Many stream ecologists have published on competition and predator–prey relationships involving mayflies. In addition to competing with other insects such as grazing caddisflies, herbivorous mayflies are known to compete for algae with snails. Benthic feeding fishes, such as darters and sculpins, are strong predators of mayfly nymphs, but these insects are also preyed upon by other insects, including predaceous stoneflies.

IV. COLLECTION AND CULTURING

The easiest way to collect mayflies is to pick up flat rocks larger than cobble from a stream and quickly look at the underside for scurrying nymphs. They can also be collected with a grab sampler or sometimes a D-net from dead leaves and other organic matter on the bottom of streams and lake littoral areas. This material can then be sorted by hand in a white tray. Some relatively large nymphs burrow into stream banks, but these are harder to find and may require some digging. Individuals or swarms of adults can be found along the banks of streams in spring, but the timing of emergence can be tricky to predict.

Mayfly nymphs can be kept in an aquarium with dead organic matter for food and clean, oxygenated water, but they will not be readily apparent to observers. You can, however, keep them in small bowls for short periods to observe their behavior.

V. REPRESENTATIVE TAXA OF MAYFLIES: ORDER EPHEMEROPTERA

A. Minnow Mayflies or Swimmers (Fig. 20a): Acanthometropodidae, Ameletidae, Ametropodidae, Baetidae, Oligoneuridae, Siphlonuridae, Metretopodidae, Isonychiidae, Arthropleidae

Identification: Streamlined, fast swimming mayflies, typically brown or striped brown and cream. **Size:** 2–8 mm. **Range:** Found throughout North America. **Habitat:** Typically in swift rivers and streams, sometimes under rocks or logs, but often trying to blend in on exposed surfaces. Rarely in ponds. **Remarks:** These are the nymphs of the adult olives, late drakes, gray drakes, pseudogray drakes, slate drakes, and brown duns. Shown here is the genus *Siphloplectron*.

Minnow Mayflies or Swimmers (Fig. 20b):

Additional Remarks: Representative of the genus *Baetis*, one of the most common and widespread genera.

Minnow Mayflies or Swimmers (Fig. 20c):

Additional Remarks: The genus *Isonychia* has long hairs on the front legs to filter food from the water.

Minnow Mayflies or Swimmers (Fig. 20d):

Additional Remarks: *Siphlonurus* is common in quiet portions of streams and rivers.

Minnow Mayflies or Swimmers (Fig. 20e):

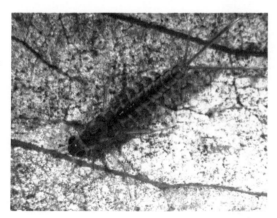

Additional Remarks: *Callibaetis* is one of the few minnow mayflies found in ponds.

Minnow Mayflies or Swimmers (Fig. 20f):

Additional Remarks: *Ameletus* has numerous undescribed species.

Minnow Mayflies or Swimmers (Fig. 20g):

Additional Remarks: *Ametropus* burrows shallowly in sand with only its eyes protruding. It creates a vortex to collect diatoms for food.

Minnow Mayflies or Swimmers (Fig. 20h):

Additional Remarks: *Acentrella* lives in erosional habitats and feeds on detritus.

B. Crawlers

Super Crawlers (Fig. 20i): Ephemerellidae: Ephemerella, Caudatella, Drunella, Serratella, Caurinella, Eurylophella, Timpanoga, Dentatella, and *Danella*

Identification: Square bodied with strong legs, sometimes with large spikes on dorsal surface or on lateral edges of abdomen. Usually brown, red, gray, or black, or striped or spotted. **Size:** 2–10 mm. **Range:** Found throughout North America. **Habitat:** Typically in swift rivers and streams, under rocks or logs or other heavy debris. Always in well-oxygenated water. **Remarks:** These are the nymphs of the sulfurs, pale-winged duns, blue-winged olives, and hendricksons. Shown here is *Drunella doddsi.*

Super Crawlers (Fig. 20j):

Additional Remarks: The genus *Drunella* contains many species that can be very large, some with horns.

Super Crawlers (Fig. 20k):

Additional Remarks: The genus *Caudatella* can have the central tail more than twice as long as the body.

Super Crawlers (Fig. 20l):

Additional Remarks: The western *Timpanogea hecuba* is broadly flattened, with lateral spines.

Super Crawlers (Fig. 20m):

Additional Remarks: The genera *Ephemerella* (shown here), *Serratella*, and *Eurylophella* are difficult to separate without a microscope.

Feeble Crawlers (Fig. 20n): *Leptophlebiidae, Neoephemeridae, Euthyplocidae*

Identification: Streamlined or flattened, rarely swimming. Brown in color or striped brown and cream. Rarely with tusks or spikes. **Size:** 2–12 mm. **Range:** Found throughout North America. **Habitat:** Typically in swift rivers and streams under rocks or logs. Some species occur in well-vegetated ponds or in lakes. **Remarks:** These are the nymphs of the black quills and the blue quills. Shown here is the genus *Paraleptophlebia*.

Feeble Crawlers (Fig. 20o):

Additional Remarks: Example of a species of *Paraleptophlebia* with tusks, which are used for digging.

Tiny Crawlers (Fig. 20p): *Caenidae, Leptohyphidae*

Identification: Flat crawling or sprawling mayflies, typically gray, brown, yellow, often with stripes. **Size:** 1–5 mm. **Range:** Found throughout North America. **Habitat:** Typically in rivers and streams, under rocks or logs or in sand. Sometimes in ponds on large debris. **Remarks:** These are the nymphs of the angler's curses and trichos. Shown here is the genus *Caenis*.

Tiny Crawlers (Fig. 20q):

Additional Remarks: Shown here is the genus *Tricorythodes*, one of the trichos.

*Armored Mayflies (Fig. 20r): **Baetiscidae:** Baetisca*

Identification: Domed, crawling mayflies, typically brown or striped brown and cream, with a single long spine on each side of the body. **Size:** 2–8 mm. **Range:** Central and eastern North America. **Habitat:** Rivers and streams, in sand with organic material, sometimes under rocks or logs, or on roots under banks and overhanging vegetation. **Remarks:** These are the only nymphs with a single lateral spine.

C. Clinging Mayflies

Clingers (Fig. 20s): Heptageniidae, Pseudironidae

Identification: Broadly flattened, with a disc-shaped head. In life colored yellow, brown, or dark red, often with darker spot patterns. **Size:** 2–15 mm. **Range:** Found throughout North America. **Habitat:** Swift, well-oxygenated rivers and streams, living under rocks or logs, sometimes trying to blend in on exposed surfaces. **Remarks:** These are the nymphs of the cahills, march browns, and quill gordons.

Clingers (Fig. 20t):

Additional Remarks: Example of a large, light-colored clinger.

Clingers (Fig. 20u):

Additional Remarks: Example of a large, dark-colored clinger.

D. Burrowing Mayflies

Burrowers (Fig. 20v): Ephemeridae, Potamanthidae, Polymitarcidae, Behningiidae, Palingeniidae

Identification: Long-bodied with feathery gills, usually with tusks for digging. Usually yellow, brown or with a pattern of stripes and spots. **Size:** 2–14 mm. **Range:** Found throughout North America, but rarer in the west. **Habitat:** Lakes, ponds, slow sections of large rivers. Burrows are in the bottom or in banks. **Remarks:** These are the nymphs of the hexes, golden drakes, white flies, and big drakes. Shown here is the genus *Dolania*.

Burrowers (Fig. 20w):

Additional Remarks: The genus *Ephemera* has tusks which are used for digging.

Burrowers (Fig. 20x):

Additional Remarks: Members of the genus *Anthopotamus* are the golden drakes.

Chapter 21

Dragonflies and Damselflies: Insect Order Odonata

I. INTRODUCTION TO THE ORDER ODONATA

Another ancient insect order is Odonata, which consists of about 318 species of dragonflies (suborder Anisoptera) and 132 species of damselflies (Zygoptera) in North America. This aquatic group has a close evolutionary link to the order Ephemeroptera but differs greatly from mayflies in body form, life history, and ecology. These predaceous insects are relatively long-lived in both the nymphal (or larval) and adult stages, but especially in the former. Odonates are relatively well known to the average person because the adults are colorful, relatively large, and easily visible as they flit about all freshwater ecosystems and nearby lands in pursuit of insect prey, including pesky mosquitoes, blackflies, and deer flies. However, they are harmless to humans—except for an occasional finger nipping if you hold them carelessly! Their bright colors and distinctive patterns are not only useful in some species identifications but also make odonates attractive to nature lovers and many artists. In fact "dragonfly watching" has become very popular, and many guides are available for identifying adults. The nymphal stages also consume many invertebrates, and even small fish in rare cases. These larvae are much more diverse in body form than the adults, especially among the dragonflies.

Odonates are predominately a tropical and subtropical group, but some species thrive in habitats as far north as Canada. About two-thirds of the species colonize pools, wetlands, and lakes, with the remaining species living in small streams through large rivers. Both adults and nymphs of a few taxa in Hawai'i are semiterrestrial. Although the distribution and abundance of many odonates have been significantly altered by habitat destruction and creation (e.g., for lentic species colonizing reservoirs and ponds), only one species has thus far been accorded federal protection: Hine's emerald dragonfly (*Somatochlora hineana*).

For more detailed classification of Odonata, see Westfall and May (1996) on damselflies, Needham et al. (2000) on dragonflies, and Merritt et al. (2008) on all aquatic insects, including a chapter on Odonata.

II. FORM AND FUNCTION

A. Anatomy and Physiology

Odonate nymphs are easy to distinguish from all other aquatic insects based on the labium, a prominent hinged mouthpart located below and at the anterior end of the head which is extended rapidly to grasp and sometimes impale prey. The shape of short to long labium ranges from spoon- or cup-shaped (especially in dragonflies) to relatively flat. In dragonflies it may resemble a "mask" covering the lower portion of the "face." The conspicuous compound eyes of odonates are especially prominent in large and highly motile hunters, while their antennae are short and somewhat filamentous.

ISBN 978-0-12-381426-5, DOI: 10.1016/B978-0-12-381426-5.00021-1

Nymphs and adults of the two suborders are simple for even amateurs to tell apart. Damselfly nymphs are long and slender and have three long, tail-like gills (caudal lamellae) which can be sac- or plate-like (one or more of the gills may be broken off). These gills aid the integumentary (skin) respiration found in all odonates. By contrast, dragonfly nymphs are usually relatively stubby and broad, with no external gills. (In dragonflies, water is moved in and out of a chamber at the end of the abdomen by muscular action to provide oxygenated water.) Dragonfly nymphs are more diverse in shape than damselfly nymphs and all adult odonates. Adult damselflies are more slender than dragonflies. When they alight on a branch, the paired, unfolded, and heavily veined (for support) wings are held to the rear and either over or to the side of the abdomen. Dragonfly wings, on the other hand, are always extended to the side and perpendicular to the long axis of the body.

B. Reproduction and Life History

Like the mayflies (Ephemeroptera) and stoneflies (Plecoptera), dragonflies and damselflies exhibit hemimetabolous development with egg, larva (nymph in Odonata), and adult stages. The moderately elaborate mating process in adult odonates has been well documented and involves copulation while the mates are flying or perched on a branch or other surface. The female may mate with multiple males, but the eggs are typically fertilized only by the last partner. When her eggs have developed sufficiently, the female deposits them by either "dipping" her abdomen into the water or depositing them above the water in vegetation or on other surfaces. Depending on the species and water temperatures, the egg stage may last anywhere from just under a week to as long as 2 months for direct development or 80–200 days in delayed (diapausing) development. After hatching, the nymph undergoes 11–13 instars, but numbers vary somewhat among and within species. The final larval stage crawls out of the water a distance of a few centimeters to a couple of meters where it molts into the adult form. One to two generations per year are most common in Odonata; but as you move toward the poles, development requires one to multiple years. Damselflies typically live a year or less, whereas North American dragonflies survive 1–6 years depending on the species, climate, and food availability. The adult stage lasts only a few weeks, with the adult life span dependent on the species and environmental conditions, as well as the presence of hungry predators and amateur or professional entomologists!

III. ECOLOGY, BEHAVIOR, AND ENVIRONMENTAL BIOLOGY

A. Habitats and Environmental Limits

Almost all species of odonates live in freshwater habitats. The exceptions are a few taxa in (a) estuarine and other inland brackish water habitats; (b) saline coastal marshes (e.g., the dragonfly *Erythrodiplax berenice*) but not in open ocean habitats; and (c) moist semiterrestrial habitats (e.g., the damselfly genus *Megalagrion* in Hawaii). Some species also thrive in unusual tropical habitats, such as tree holes and other aquatic habitats formed by plants. While some taxa are restricted to flowing water, such as the dragonfly family Cordulegastridae, odonates in general seem more common in shallow near-shore waters of lentic environments where their abundance is partially correlated with the amount of vegetation and organic detritus.

The shape of a nymph's legs and body often closely match the organism's habitat. For example, nymphs of damselfly and dragonfly species with elongated bodies, such as hawker dragonflies in the family Aeshnidae, can be most often found clinging to vegetation. Species with broad flattened bodies, such as skimmers or perchers in the dragonfly family Libellulidae and clubtail dragonflies in the family Gomphidae, reside mostly on or in the substrate, respectively. Sometimes odonate taxa are grouped by how they occupy and move through habitats; these include more sedentary burrowers and sprawlers and more active climbers and crawlers. These groupings have a weak correlation with taxonomy, but they provide hints about the animal's ecology as well as its body and leg structure. An odonate's residence also influences the kind of prey it consumes.

B. Functional Roles in the Ecosystem and Biotic Interactions

All odonates are predators, but they vary in how they find their prey (stalking or sit-and-wait predators) and their hunting grounds (on the bottom or up in vegetation), both of which affect the kind of prey encountered and captured. Prey abundance influences choice of hunting grounds in some species. Some damselflies can swim weakly, but this is more a means to escape predators or to reach more desirable sites rather than a hunting technique. Visual acuity, including eye size, is strongly correlated with tactics for encountering prey. Insects are the most common prey type, but odonate nymphs will also eat almost any organism they can capture and physically consume, including benthic and planktonic invertebrates. Large dragonflies, like aeshnids, will occasionally capture small fish and tadpoles. These and other larger prey items cannot be swallowed whole, so the dragonfly will eat them piecemeal. Based on their abundance, odonates are often the apex predators in ecosystems lacking fish.

While patterns of microhabitat use and mobility influence prey capture, they also reflect chances of predation on odonates. One of the most dangerous predators to an odonate is another odonate! But, they are also attacked by various vertebrates (especially fish) and invertebrates, such as crayfish and some predaceous bugs and beetles. When threatened, an odonate will attempt to hide or flee by crawling, jetting (using a forceful expulsion of water from the rectal chamber of dragonflies), or swimming (damselflies using undulatory movements their abdomen and caudal lamellae). Once attacked, damselflies are mostly unable to resist unless they can flee. Dragonflies, however, rely partially for protection on dorsal abdominal spines. Some evidence suggests that spine length correlates with predator abundance, much as do body shape and spine length in some water fleas (Branchiopoda).

IV. COLLECTION AND CULTURING

Odonates are relatively easy to collect with an aquatic D-net for nymphs or a sweep net for adults. Both suborders of nymphs are relatively slow moving, as are adult damselflies, but adult dragonflies require a bit more skill to capture because some are exceedingly fast flyers. The easiest time to collect adult dragonflies is in the morning when air temperatures are cool enough to keep the dragonflies from flying until the sun heats their muscles. Adult damselflies can be captured on the wing or when they perch on a branch. Look for nymphs among aquatic plants, clinging to wood snags, hiding in organic debris, or under the margins of rocks. Some species have more unique habitat requirements, such as some dragonflies in the family Gomphidae which tend to bury themselves in mud or sand with only an extended abdomen above the bottom to obtain oxygenated water.

Odonates are easily maintained in aquaria as long as live insects or worm prey are available and predaceous fish are absent.

V. REPRESENTATIVE TAXA OF DRAGONFLIES AND DAMSELFLIES: ORDER ODONATA

A. Darners (Fig. 21a): Aeshnidae

Identification: Streamlined, often bright green or brown dragonfly nymphs, with the large labial mouth parts not covering the "face." The legs are slender, the eyes are very large, and the body is kept clean. **Size:** 2–65 mm. **Range:** Found throughout North America. **Habitat:** Occurs in vegetation and on logs, roots, or branches in lakes, ponds, or quiet backwaters or rivers and streams. **Remarks:** These are the nymphs of the adult darners, dragonhawks, and hawkers. These are the largest of the dragonflies.

B. Petaltails and Clubtails (Fig. 21b): Petaluridae, Gomphidae

Identification: Squat, rough, sometimes flattened dragonfly nymphs, with the large labial mouth parts not covering the "face." The legs are stiff, the eyes are not especially large, and the body is often covered in debris for camouflage. **Size:** 2–50 mm. **Range:** Found throughout North America. **Habitat:** Petaltails occur is springs and seeps while clubtails occur in springs, rivers, and streams. Some clubtails burrow in clean sand or in loose organic debris, and may have an elongated abdomen segment to act as a "snorkel." **Remarks:** The petaltails are an ancient family, with fossil forms from the Jurassic.

Petaltails and Clubtails (Fig. 21c):

Additional Remarks: Some clubtails may be remarkably flattened, for hiding in sand or debris, while waiting to capture passing prey.

C. Spiketails (Fig. 21d): Cordulegastridae

Identification: Long-bodied, brown dragonfly nymphs, with the large labial mouth parts extending over the "face." The "jaws" of the labium with large, obvious jagged "teeth." The eyes are small and raised above the head. **Size:** 2–45 mm. **Range:** Found throughout North America. **Habitat:** Typically lying buried in debris, mud, or sand, or sometimes under rocks in forested areas. **Remarks:** These nymphs do not dig with their front legs. Instead they dig with their hind legs and squirm down into the substrate.

D. Skimmers, Cruisers, and Emeralds (Fig. 21e): Libellulidae

Identification: Squat, sprawling, sometimes hairy. Usually mottled with browns, yellows, black, and gray. The large labial mouth parts extend up to cover the lower portion of the "face." The body is often coated with algae or debris. **Size:** 2–35 mm. **Range:** Found throughout North America. **Habitat:** Occurs in vegetation near the bottom, or lies buried in debris. **Remarks:** This is the largest family of dragonflies with close to 250 species in North America.

Skimmers, Cruisers, and Emeralds (Fig. 21f):

Additional Remarks: Members of the subfamily Macromiinae tend to have very long legs.

E. Slender Damselflies (Fig. 21g): Calopterygidae, Lestidae

Identification: Very slender or elongated damselfly nymphs, ranging in color from brown to green or white. Calopterygids (figured) have thick straight antennae, while lestids have thin, slender antennae. **Size:** 2–40 mm. **Range:** Found throughout North America. **Habitat:** Occurs in vegetation in lakes, ponds, or quiet backwaters or rivers and streams. **Remarks:** These are the nymphs of the adult rubyspots, jewelwings, and spreadwing damselflies.

F. Common Damselflies (Fig. 21h): Coenagrionidae, Protoneuridae

Identification: Damselfly nymphs ranging in color from brown, yellow, gray, green or white, and having thin, slender antennae. **Size:** 2–30 mm. **Range:** Coenagrionids are found throughout North America and are the most common damselflies. Protoneurids are limited to southern Texas, Mexico, and the Antilles Islands. **Habitat:** Occurs in vegetation in lakes, ponds, or quiet backwaters or rivers and streams. **Remarks:** In parts of Asia, these animals are symbols of victory and joy.

Common Damselflies (Fig. 21i):

Additional Remarks: These are the nymphs of the bluets, red damsels, violet dancers, vivid dancers, sprites, and forktails.

Chapter 22

Stoneflies: Insect Order Plecoptera

I. INTRODUCTION TO THE ORDER PLECOPTERA

Stoneflies are another small, hemimetabolous order of insects whose nymphs are entirely aquatic and mostly found in streams. About 660 species have been described from North America, but on average about five new species are described each year. Species richness tends to peak in mountainous ecoregions and declines precipitously in the lowland and turbid streams of the Great Plains where stream beds are often sandy and solid substrate is rare.

None of the Plecoptera are harmful to humans in any life stage, and adults rarely reach densities where they become a nuisance (unlike mayflies), and then in only a few species living in rivers, such as the genus *Perlesta* (family Perlidae).

As is the case for most insects, stoneflies are rarely if ever included on lists of threatened or endangered species, even though extinction events are not unknown. Some plecopterans, however, are considered "species at risk" in states and provinces of the USA and Canada because of habitat destruction and excessive pollution or sedimentation. Stoneflies are more vulnerable to extinction than some other insect groups because adults are weak flyers and cannot disperse far in search of new habitats. Some scientists have suggested that increased precipitation variability is causing a gradual change in species dominance within plecopterans of the Midwest as species with shorter life cycles and the ability to enter diapause replace formerly dominant taxa lacking those characteristics.

II. FORM AND FUNCTION

A. Anatomy and Physiology

Stonefly nymphs are typically 5–70 mm long and can be distinguished from other insects by the combined presence of two stout tail-like cerci (Fig. 22a), a pair of long antennae, and three-segmented legs with paired claws (Fig. 22b). They have a moderately prominent pair of compound eyes along with several eyespots (ocelli). Two pairs of wing pads are present on the thorax, at least in older nymphs. Gills are most commonly found on the ventral side of the thorax, but these may occur in some species on the head and abdomen. Students and amateur naturalists are most likely to confuse stoneflies with mayfly nymphs, especially if one of the three cerci of the mayfly has been lost. In those cases, look for the distinctive paired claws on all six legs which characterize plecopterans.

Adult plecopterans somewhat resemble the nymphs, at least in general body shape. However, they have long net-veined wings which extend past the abdomen and are held flat over the abdomen when the stonefly is at rest. The hind wings have a section that folds under the main wing, which accounts for the ordinal name meaning "folded wing."

ISBN 978-0-12-381426-5, DOI: 10.1016/B978-0-12-381426-5.00022-3

FIG. 22A. The stonefly nymph *Taeniopteryx*. Note the two stout, tail-like cerci.

FIG. 22B. The stonefly nymph *Pteronarcys*. Note the three-segmented legs with paired claws.

B. Reproduction and Life History

The life cycle of plecopterans varies from 4 months to 3–4 years, depending on the species, location (latitude and altitude), and general environmental conditions. No North American species produces more than one generation per year. These cycles are generally classified as univoltine (requiring 1 year from egg to adult) or semivoltine (really meaning 2 years, but used for all multi-year cycles). "Fast univoltine cycles" usually involve a egg diapause of up to 8 months followed by a rapid 4–7 months of development to produce the reproducing adult stage. Emergence of nymphs from eggs in these species usually occurs in spring or summer. Other fast univoltine species undergo nymphal diapause (often in the hyporheic zone) and emerge as adults in December through March (winter stoneflies). In slow univoltine species, the nymph has an extended non-diapausing development followed by emergence of the adult, all within a year. Semivoltine life cycles generally occur where temperatures are low throughout most of the year, as in high latitudes and altitudes. Adults of most taxa with predaceous nymphs emerge as adults in the spring or summer.

Stoneflies pass through 12–24 instars during development from a first instar nymph to an adult capable of reproduction. At the end of their developmental period, the stonefly crawls from the water onto a hard substrate to complete development. Adults live about 1–4 weeks and usually feed on nectar, algae, lichens, soft fruit, or other organic matter, if they consume any nutrition during this stage. Winter stonefly adults tend to live longer. Adults attract mates with species-specific drumming patterns produced by tapping the tip of the abdomen on a surface.

   A bizarre exception to this life history pattern occurs in the stonefly *Utacapnia tahoensis* (family Capniidae). This species lives far below the surface of the very deep Lake Tahoe and resides within water through its entire life cycle rather than having the otherwise omnipresent terrestrial adult phase.

III. ECOLOGY, BEHAVIOR, AND ENVIRONMENTAL BIOLOGY

A. Habitats and Environmental Limits

Unlike other orders of insects, stoneflies are almost entirely restricted to permanent streams. A few species live in wave-swept and well-oxygenated shorelines of cold oligotrophic (low productivity) lakes of alpine and northern regions, and some taxa are reported from intermittent streams. Stoneflies also colonize the hyporheic zone of river gravel beds where they have been collected from temporary wells dug as far as 3 km away from the river's surface channels. To exist in such habitats and complete an entire life cycle, the stonefly needs a steady flow of nutrients from the main river and must be able to reach the stream surface in order to emerge as an adult.

   Adults typically reside very near their home stream because they are poor flyers and have an adult phase lasting only one to a few weeks.

   Stoneflies are most often found in cool streams with mostly low to sometimes moderate pollution levels. They represent a third of the common environmental quality index of species composition known as "EPT" (Ephemeroptera, Plecoptera, and Trichoptera). From this you might assume that they are intolerant of pollution, but in fact some species tolerate somewhat poor water quality conditions even though most taxa are restricted to clean, silt-free, and well-oxygenated streams. Where oxygen is low, some stonefly species flex their body up and down (as in the family Perlidae) or swing the abdomen from side to side (as in the giant stoneflies of Pteronarcyidae) to increase the flow of oxygen to their gills.

B. Functional Roles in the Ecosystem and Biotic Interactions

Stoneflies can be very abundant in rocky streams, where they may represent as much as 10% of the invertebrate production. As you would surmise, therefore, they play pivotal roles in stream food webs. As evidenced by anatomical differences in nymph mouthparts, stoneflies are adapted for either herbivory (mostly shredding, but also scraping, gouging, and general gathering) or predation, but the functional feeding group can vary as the nymph matures. Members of the stonefly families Capniidae (often called "winter stoneflies") and Pteronarcidae are especially known for their abilities as shredders. Predatory stoneflies tend to be mobile hunters rather than sit-and-wait predators; and in small streams, they may be the top invertebrate predator. Common prey include chironomid midges, mayflies, caddisflies, and even small stoneflies.

   Stoneflies are subject to predation from larger invertebrate predators (e.g., hellgrammites) and fish. They commonly crawl upon the substrate, especially at night; but if forced to swim to flee a predator or when knocked off the substrate by currents, the nymphs will swim weakly with side-to-side abdominal movements.

IV. COLLECTION AND CULTURING

The most reliable places to collect stoneflies are small rocky streams, especially those in more mountainous regions. In almost any stream, however, they can be found if you look hard enough. If the stream is sandy, seek stoneflies on wood snags or around the hard surfaces added to prevent stream erosion. In rocky streams, search the surfaces of cobble (especially the underside), use a kick net to catch insects stirred up by your boots, employ a small net to collect from the surface of snags, or collect aged leaf packs for the hidden stoneflies.

   Maintaining stoneflies in an aquarium requires cool temperatures, plenty of oxygen, and sufficient food. The last is trickier because you need to identify the stonefly well enough to know whether it eats dead organic matter (colonized with bacteria and fungi), benthic algae, or other insects.

V. REPRESENTATIVE TAXA OF STONEFLIES: ORDER PLECOPTERA

A. Salmonflies (Fig. 22c): Pteronarcyidae

Identification: Large dark brown to black stoneflies, sometimes with yellow longitudinal stripes. Multibranched gills under the head, at the bases of the legs, and under the first couple abdominal segments. **Size:** 10–55 mm. **Range:** Widespread in North America. **Habitat:** Rivers and streams. **Remarks:** Fly fishing patterns established on salmonfly nymphs include brown nymphs and dorsata nymphs.

B. Roachflies (Fig. 22d): Peltoperlidae

Identification: Small to medium dark brown stoneflies, most of which are squat and sprawling, giving the appearance of a cockroach. A gill with a single (rarely two) branch occurs at the base of each leg. **Size:** 5–20 mm. **Range:** Mountainous regions of western and eastern North America. **Habitat:** In rivers and streams, in leaf litter and leaf pack, or under logs or in woody debris jams. **Remarks:** Unlike most stoneflies, which are predators, these stoneflies are leaf shredders.

C. Springflies and Stripetails (Fig. 22e): Perlodidae

Identification: Small to medium stoneflies, brown, yellow, green, or black, sometimes with yellow longitudinal stripes, but always patterned with several colors. Obvious gills are usually absent (one species has long gills along the abdomen), or a single gill is present at each leg. **Size:** 10–55 mm. **Range:** Widespread in North America. **Habitat:** Rivers and streams. **Remarks:** These are the nymphs of the medium browns, willow flies, and yellow stones.

D. Stones (Fig. 22f): Perlidae

Identification: Medium-sized stoneflies, often patterned with strongly contrasting yellow and black, or may be dull browns and grays, sometimes flat black. Multibranched gills are present at the bases of the legs, and sometimes at the end of the abdomen. **Size:** 10–30 mm. **Range:** Widespread in North America. **Habitat:** Rivers and streams. **Remarks:** These are the nymphs of the golden stones, pale stones, keel stones, and brown willow flies.

E. Sallflies (Fig. 22g): Chloroperlidae

Identification: Small, slender, yellow green or brown stoneflies, lacking gills. Sometimes patterned with dorsal spots, or a medial yellow stripe. **Size:** 10–55 mm. **Range:** Widespread in North America. **Habitat:** Rivers and streams. These species tend to burrow in sand and gravel, sometimes fairly deep into the substrate. **Remarks:** These are the nymphs of the little greens, little yellows, sallies, and yellow sallies.

F. Winter Stoneflies, Snow Stoneflies, Needleflies (Fig. 22h): Capniidae, Leuctridae, Taeniopterygidae

Identification: Small, black, tan, brown stoneflies, sometimes with yellow transverse stripes. Gills are lacking. Wing pads are either parallel or divergent. **Size:** 5–10 mm, very rarely to 20 mm. The legs are shorter than the abdomen. **Range:** Widespread in North America. **Habitat:** Rivers and streams. **Remarks:** In most species the adults emerge in the winter.

G. Forestflies, Mottled Browns (Fig. 22i): Nemouridae

Identification: Small, brown stoneflies, sometimes with the wing pads diverging. Gills may or may not be present. If present, they are either single or multi-branched under the head. The legs are as long as or longer than the abdomen. **Size:** 5–8 mm. **Range:** Widespread in North American forested regions. **Habitat:** Rivers and streams. **Remarks:** The adults of some species may emerge in winter.

Chapter 23

True Bugs: Insect Order Hemiptera

I. INTRODUCTION TO THE ORDER HEMIPTERA

Contrary to the vernacular spoken routinely by the average person on the street—and even casually by most aquatic ecologists—all insects are not bugs! Indeed, there is a large taxonomic group of true bugs, but the name and taxonomic relationships of this group are frequently disputed by entomologists. In that regard, we follow here recommendations in the Tree of Life project (http://tolweb.org/Hemiptera) to classify true bugs as the suborder Heteroptera in the order Hemiptera. This order contains a diverse mixture of groups, including aphids, cicadas, and true bugs. All freshwater hemipterans are also heteropterans, and all share the ordinal feature of having specialized elongated mouthparts useful for sucking fluids from other organisms. Through the remainder of this chapter "aquatic bugs" or just "bugs" will refer to freshwater members of the suborder Heteroptera.

Of the roughly 38,000 described species of heteropterans around the world, a little under 9% are aquatic and have nymphs and adults that live in the water (the majority) or on its surface, usually in nonflowing habitats. The surface-dwelling species are sometimes categorized as "semiaquatic." At least 412 species of aquatic bugs have been described from North America, but relatively little attention has been given to this group over the last half century so many other species may exist. Adults are fairly easy to identify even in the field, but classifying nymphs is more challenging even at the generic level.

Aquatic bugs are harmless unless picked up, but some species can produce a temporary but painful "bite" when their proboscis pierces an incautiously grasping hand—which has happened to the authors more than once! Some biting bugs, such as backswimmers (Notonectidae) and giant water bugs (Belostomatidae) are occasionally pests in swimming pools when accidentally encountering humans. From a more serious economic standpoint, members of the belostomatid genus *Lethocerus*, sometimes called "toe bitters," can wreak havoc in fish hatcheries. These and other large bugs will also kill small fish in home aquaria. The reason for this occasional negative impact on humans is that nymphs and adults of many species are predators. This feeding mode also benefits humans because larval mosquitoes and other biting flies are preyed upon by aquatic bugs. Humans also benefit from aquatic bugs because various species are considered a food delicacy in some parts of the world, including Mexico and some Asian countries, and some taxa like water boatmen (Corixidae; Fig. 23a) are marketed as dry food for pet turtles and aquarium fish.

The presence of large numbers of aquatic bugs is sometimes an indicator of water pollution. Although they are not necessarily more resistant to toxins than other insects, they avoid problems encountered by other aquatic insects under polluted, low oxygen levels because most bugs obtain their oxygen directly from the atmosphere rather than from dissolved supplies.

ISBN 978-0-12-381426-5, DOI: 10.1016/B978-0-12-381426-5.00023-5

FIG. 23A. A genus of water boatmen, *Hesperocorixa*.

II. FORM AND FUNCTION

A. Anatomy and Physiology

Aquatic bugs can be identified by their highly modified mouthparts which are shaped into a tube-like rostrum used to suck liquified prey tissue into the bug's body. This rostrum projects down, below, and behind the head. In addition, the overlapping fore wings of adults are hardened at the basal half and membranous closer to the tips. The shorter hind wings of the adults are entirely membranous and held beneath the larger hind wings and over the abdomen. Nymphs have partially developed wings, called wing pads. There are some differences between the leg structure of nymphs and adults, but these are not easily seen without a microscope. In some species, one or more pairs of legs have paired terminal claws. The body shape varies dramatically among species from flattened ovals (e.g., creeping water bug, Naucoridae) to slender and highly elongated shapes (e.g., water scorpions, Nepidae). Most are in the size range of 20–60 mm, but some giant water bugs reach 100 mm!

The strategies of aquatic bugs for obtaining oxygen differ markedly from other aquatic insects. All species of bugs lack external gills and rely instead on internal respiratory tubes (tracheae). When oxygen is needed by surface-dwelling species, they need only open their spiracles covering the entrance to the tracheae. Submerged species, however, follow one of two strategies. Most species travel to the surface and either expose their tracheae to the atmosphere or stick a breathing tube through the surface film, much like the snorkel used by skin divers. Other bugs carry a bubble of air on their ventral side which provides oxygen to the tracheae. Either this bubble is replenished occasionally from the water surface or, in less active species, oxygen diffuses from the water into the bubble as levels drop creating a diffusion gradient in favor of the bug.

B. Reproduction and Life History

The typical 1-year life history pattern for aquatic bugs involves egg laying in the spring, nymph development through four to five instars to adults during warmer months of summer and early fall, overwintering as adults, and mating in the late winter or spring. Eggs are secured on aquatic plants or other solid objects within the water or are placed just above the water by some semiaquatic bugs. The average time spent as a nymph is about 2 months, but ranges from 1 to 8 months depending on the species and location. In more rigorous environments, such as in northern latitudes, adults of semiaquatic species may hibernate in protective microhabitats of mud or leaves. (Hibernation differs from the diapause of other species in that full development has already occurred in the former state.) Dispersal among habitats occurs in the adult phase, but bugs usually disperse only when local conditions deteriorate. This three-stage, paurometabolous cycle (egg, larva/nymph, and adult) differs from the three-stage hemimetabolous cycle of mayflies, dragonflies, and stoneflies in that the adults and nymphs of bugs differ very little in structure and they live in the same environment.

III. ECOLOGY, BEHAVIOR, AND ENVIRONMENTAL BIOLOGY

A. Habitats and Environmental Limits

The morphology and habits of aquatic bugs make them less suited for turbulent habitats, and thus most species occur in calmer waters of ephemeral pools through shorelines of large lakes. They also occur in some salt lakes and ponds. Few species venture far from this habitat. Those found in creeks through large rivers occur almost exclusively in pools, lateral slack waters, and/or among protected aquatic weeds. Nymphs and adults colonize very similar habitats and are typically intermixed with each other.

Most aquatic bugs climb, swim, or "row" (e.g., water boatmen and back-swimmers) among aquatic vegetation in lentic and lotic habitats, probably because of food availability and reduced predation. Examples of these are giant water bugs, creeping water bugs, and water scorpions. The semiaquatic bugs, such as water striders (Gerridae; Fig. 23b) and water treaders (Mesoveliidae), skate across the surface on hydrophobic legs in search of trapped prey or insects within diving distance. Finally, a few taxa, such as water boatmen seek food near the undersurface of the air–water interface.

Although the coverage of this book is confined to inland waters, it is interesting to note that some species of water striders live on the surface of the open ocean in the absence of all other insects!

FIG. 23B. A member of the water strider family Gerridae.

B. Functional Roles in the Ecosystem and Biotic Interactions

The adults and nymphs of many species of aquatic bugs are predators, with most species feeding on a variety of invertebrate prey including mosquito larvae and aquatic bugs of the same or different species. Large species will attack larval fish of most species and the adults of smaller fish taxa like minnows. Giant water bugs have also been observed successfully attacking tadpoles, small frogs, salamanders, and crustaceans of modest size. The fore legs of bugs are generally equipped to grasp and hold struggling prey. The hind legs may be used in propulsion or for grasping a stable surface while waiting for a passing prey. Once the prey has been secured, the rostrum injects toxic enzymes which first paralyze and then digest the prey (within the prey's body) before the liquified contents are sucked into the bug's digestive tract for further digestion. This technique allows the bug to kill an organism up to 50 times its size in some cases.

Water boatmen are one of the few aquatic bugs that include some nonpredaceous species in the family. Many of these common bugs are instead collector-gatherers, feeding on algae and detritus in suspension or at the water surface.

Like many terrestrial species, such as the "stink bug," many aquatic bugs have scent glands which seem to repel some predators. Those bugs lacking this deterrent, such as water boatmen, are subject to more intense predation from some fish but especially from resident and migratory birds.

C. Stridulation

Many male true bugs stridulate. This is accomplished by rubbing together a specialized row or small ridges on one surface (called a strigil) and one or more rows on another surface. The result is a squeaking sound. Stridulation is used to attract mates. In the water boatmen the strigil is on the abdomen, whereas in the backswimmers it is on the front legs. Sometimes these animals will stridulate when they become alarmed, so do not be surprised if you capture a true bug and it begins squeaking in protest!

IV. COLLECTION AND CULTURING

Aquatic bugs can be collected from lentic or sometimes lotic habitats. In the former, the bugs can be found in submerged vegetation throughout the wetlands and near the shoreline among weeds of ponds and lakes. You can also collect them from the water surface of these habitats. In streams, aquatic bugs occur in sluggish waters, including stream pools and lateral channels with minimal or zero currents, or in thick vegetation in slow flowing areas of river swamps. When seeking fully aquatic species, use a D-net to scoop up rooted vegetation and place it in an observation pan. If you are not wearing gloves, watch out for hidden bugs that will strike if cornered while you are scooping or sorting vegetation. When seeking semiaquatic, surface bugs, a quick pounce and sweep with a D-net or even a stout butterfly net will work (though it is easy to tear the more delicate mesh). A problem for the human collector is that semiaquatic species are very wary and are watchful for both bird and fish predators, so it can be difficult to get one with a single sweep.

Maintaining aquatic bugs in an aquarium is relatively easy as long as you provide submerged physical structure, adequate live prey, a habitat lacking fish predators, and an escape-proof aquarium (nymphs crawl or swim very well and adults can fly). Getting adequate prey and making the aquarium "de-bug proof" will be the hardest tasks.

If you want to preserve aquatic bugs you can either put them in an tightly capped jar with alcohol (see Chapter 2) or you can dry some specimens of mostly semiaquatic species (or the larger submerged species) following techniques you would employ for terrestrial insect specimens.

V. REPRESENTATIVE TAXA OF TRUE BUGS: ORDER HEMIPTERA

A. Shore Bugs (Fig. 23c): Saldidae, Ochteridae

Identification: Oval dark brown to black bugs, sometimes with white speckles or mottling. Often running or jumping. Very quick to fly away. **Size:** 4–7 mm. **Range:** Widespread in North America outside of the arctic regions. **Habitat:** Shorelines of permanent or temporary habitats, sometimes on rocks, mud, or sand. Never on open water. **Remarks:** These insects are predators and scavengers, and they can leap several times their own body length.

B. Toad Bugs (Fig. 23d): Gelastocoridae

Identification: Round, flat bodied, jumping bugs. Gray, brown, or black with mottles of white or gray. **Size:** 4–9 mm. **Range:** Widespread in North America outside of the arctic regions. **Habitat:** Shorelines of permanent or temporary habitats, sometimes on rocks, mud, or sand. Never on open water. **Remarks:** These insects are predators. Some species burrow in sand.

C. Water Treaders, Velvet Water Bugs, Riffle Bugs, Water Crickets (Fig. 23e): Hebridae, Mesoveliidae, Macroveliidae, Veliidae

Identification: Oblong to triangular dark brown to black bugs, sometimes with white margins, or lighter abdomens. Often occurring in very large groups. **Size:** 3–7 mm. **Range:** Widespread in North America south of the arctic regions. **Habitat:** On the water surface at or near shorelines, sometimes on rocks, or on vegetation of permanent habitats. **Remarks:** These insects are predators or scavengers. Some species are flightless.

D. Water Measurers (Fig. 23f): Hydrometridae

Identification: Long thin, stick or hairlike insects, with thread like legs. Light brown to beige. **Size:** 4–11 mm. **Range:** Widespread in North America. **Habitat:** In quiet permanent water, in well-vegetated areas, or under overhanging banks. **Remarks:** These insects are predators. Most species lack wings or have reduced winged forms.

E. Water Striders, Water Skippers (Fig. 23g): Gerridae

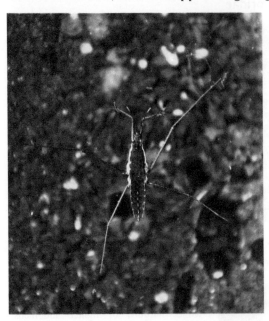

Identification: Elongated or round bodied, dark brown to black bugs, with a white or silver underside. The middle and hind leg pairs are elongated. **Size:** 4–20 mm. **Range:** Widespread in North America outside of the arctic regions. **Habitat:** Skating along on the open water surface in any quiet or slow-flowing waters. **Remarks:** These insects are predators and scavengers. Some species are wingless.

F. Pygmy Backswimmers (Fig. 23h): Pleidae

Identification: Humpbacked, round bodied, upside down swimming bugs, with the hind leg pair elongated for swimming. Color is brown to cream with brown patterning. **Size:** 1.9–2.25 mm. **Range:** North America east of the continental divide in warmer regions, as far north as Ontario and Quebec, and extending west and north into Idaho and Manitoba. **Habitat:** Well-vegetated ponds, lakes, and rivers. **Remarks:** These insects are predators.

G. Backswimmers (Fig. 23i): Notonectidae

Identification: Streamlined, cylindrical or round bodied, upside down swimming bugs, with the hind leg pair elongated for swimming. The underside is black or white, but the dorsal surface may be black, white, yellow, or crimson. **Size:** 5–20 mm. **Range:** Widespread in North America. **Habitat:** Any quiet or slow-flowing aquatic habitat. **Remarks:** These insects are predators and can give a sharp bite. They will sometimes squeak.

H. Boatmen, Water Boatmen, Oarsmen (Fig. 23j): Corixidae

Identification: Streamlined, flat bodied, right side up swimming bugs, with the hind leg pair elongated for swimming. The underside is cream or white, sometimes black. The dorsal surface may be black, white, gray, or yellow with various markings. **Size:** 5–20 mm. **Range:** Widespread in North America. **Habitat:** Any quiet or slow-flowing aquatic habitat. **Remarks:** These insects are herbivores and will not bite. They will sometimes squeak.

I. Creeping Waterbugs, Toebiters (Fig. 23k): Naucoridae

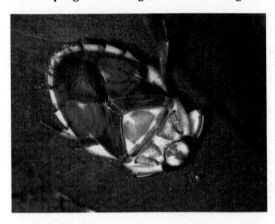

Identification: Round, flat bodied, swimming bugs, with sharp, rigid front legs. May be green, brown, or yellow. **Size:** 4–10 mm. **Range:** Widespread in North America south of the arctic and subarctic regions. **Habitat:** Permanent ponds, lakes, streams, and rivers. Typically in vegetation or algae. **Remarks:** These insects are predators and can give a painful bite.

J. Giant Waterbugs, Toebiters, Electric Light Bugs (Fig. 23l): Belostomatidae

Identification: Large, flat bodied, swimming bugs, with strong front legs. Always brown. Females often lay their eggs on the male's back. **Size:** 12–60 mm. **Range:** Widespread in temperate and subtropical North America. **Habitat:** Permanent or temporary ponds and lakes. Typically in vegetation, rarely in open water. **Remarks:** These insects are predators and can give a painful bite. They sometimes come to lights in large numbers.

K. Water Scorpions (Fig. 23m): Nepidae

Identification: Large, flat or cylindrical, elongated bugs, with long legs and a pair of long breathing tubes projecting from the tail. Always brown. Poor swimmers. **Size:** 12–100 mm. **Range:** Widespread in temperate and subtropical North America. **Habitat:** Permanent or temporary ponds and lakes, slow streams, and springs. Typically in vegetation. **Remarks:** These insects are predators and can give a painful bite. They are poor swimmers and rely on camouflage to capture their prey.

Chapter 24

Hellgrammites, Spongillaflies, Caterpillars, and Others: Minor Aquatic Insect Orders

I. INTRODUCTION TO THREE ORDERS

This chapter is a bit of a grab bag because it includes three insect orders that are either small in general species richness or specifically limited in numbers of aquatic species: Megaloptera (hellgrammites), Neuroptera (spongillaflies), and Lepidoptera (aquatic caterpillars). Of these orders, only the megalopterans are both commonly encountered and large enough to be easily noticed. All three orders feature holometabolous development (egg, larva, pupa, and imago/adult). Adult and most pupal stages are terrestrial.

Megaloptera is a small insect order with less than 400 species described worldwide; it was previously included within the order Neuroptera. They are divided into two families: Corydalidae (fishflies and larval hellgrammites/adult dobsonflies) with 22 species described for North America and Sialidae (alderflies) with 24 species. Many members of this order are very large (up to 90 mm), and all are predators of other aquatic invertebrates. Megalopterans are entirely harmless to humans unless carelessly held, in which case the strong larval mandibles can deliver a sharp bite. Corydalids are most often associated with well-oxygenated, clean waters, while sialids typically live in habitats with somewhat lower oxygen conditions and may thrive in substantially impaired environments with organic wastes or chemical pollutants.

Neuroptera is a moderately small order of mostly terrestrial species, such as antlions and lacewings, but the only family in North America that is entirely aquatic is Sisyridae, with a mere six species of spongillaflies. The common name of this family reflects the observations that these organisms live within and feed upon freshwater sponges found in permanent lotic and lentic habitats. The limitation of sponges to clean water also means that spongillaflies do not occur in polluted waters. Spongillaflies do not harm or even interact with humans except for an occasional invertebrate zoologist!

The order Lepidoptera is known to almost everyone in the form of moths and butterflies, but it surprises most people to learn that the order includes about 50 aquatic species in North America, most within the family Pyralidae. Those considered aquatic have larval and often pupal stages that live in water, primarily on rooted aquatic plants which extend to the surface or float upon it. Most aquatic caterpillars are associated with wetlands, ponds, lakes, and backwaters of large rivers where abundant floating vegetation is present, but some species also occur in flowing water on rocks. They typically graze on algae and external vascular plant tissue or burrow into the plant. Only rarely do they exert a strong control of aquatic plant populations, but they have been investigated as a possible biological control agent for invasive plants, such as water hyacinths. They can survive in any habitat where aquatic macrophytes grow in freshwaters.

ISBN 978-0-12-381426-5, DOI: 10.1016/B978-0-12-381426-5.00024-7

II. FORM AND FUNCTION

A. Anatomy and Physiology

Megaloptera: Members of this order are easily distinguished from other insects, except for some species of larval beetles that also possess abdominal filaments. Larvae have broad, heavily sclerotized heads with massive mandibles (especially in Corydalidae). The thorax is also hardened dorsally but is soft ventrally. It contains six legs, each with a pair of terminal claws. Wing pads are absent. Other than the massive jaws, the most conspicuous features are seven to eight pairs of long, tapering, and segmented filaments projecting laterally from the abdomen. The abdomen terminates either in a pair of prolegs each with two hooks (Corydalidae) or in a single long tapering filament lacking claws (Sialidae). Corydalid larvae are larger (on average 30–65 mm in the last larval stage) than sialid larvae (25 mm).

Adults are large with prominent mandibles and net-veined wings which are held tilted (roof-like) over and to the side of the abdomen when at rest. The hind wings are larger than the fore wings and pleated to fold under the fore wings when not in use. Adults are relatively weak fliers.

Few megalopterans have gills and instead absorb oxygen while underwater through the general body wall and the thin abdominal filaments. In a few species living in shallow lentic environments, oxygen is obtained directly through a pair of long abdominal tubes which are projected into the atmosphere. The usual absence of gills confines most megalopterans to well-oxygenated habitats of streams and lake shorelines.

Neuroptera: Larval spongillaflies are small (4–8 mm), stout-bodied insects with abundant spiny setae. Most larvae are yellow or green, which matches well with the sponge habitat, while the adults have lacy wings and dull-colored bodies. Their mouthparts consist of a pair of long unsegmented stylets adapted for obtaining soft tissue from the sponge body. Other than these feeding adaptations, the larva has no prominent adaptations for life in water. Not surprising, therefore, they obtain oxygen through their body wall and segmented abdominal gills. Flexing movement of the body helps ensure a steady flow of oxygenated water across the outer body wall.

Lepidoptera: Aquatic moths or caterpillars are easily distinguished from other aquatic insects but are readily confused with other terrestrial moths that may be present on the upper side of floating aquatic plants. The most distinguishing larval characteristic is the presence of pairs of short prolegs ringed with hook-like crochets on most but not all abdominal segments. Conspicuous breathing spiracles occur on one thoracic segments and most abdominal segments. Profuse filamentous gills cover the bodies of some species, but most other taxa respire directly across the soft outer body wall. At least one species uses hydrophobic hairs to hold a bubble of air while the larva is submerged.

B. Reproduction and Life History

Megalopteran species are longer lived than many other insects, with life cycles averaging 2–3 years but ranging from 1 to 5 years, depending particularly on the climate and species. Members of this order do not undergo diapause, and pass through 10–11 instars. Larvae that live in intermittent streams must reside in isolated pools to survive during low water periods. At the end of the larval period, these insects crawl on to land to find a safe place to pupate. Adults typically emerge at night and are secretive in habitat. Their brief lives involve searching for mates, copulation, and egg laying in masses attached to objects above the water. They rarely feed (possibly some liquid food) and have short lives (about 3 days for males and 8–10 days for females).

Spongillaflies appear to produce multiple generations per year, with each larva undergoing three instars. The final stage crawls out of the water to find a protected pupation site within a few meters of the water and spins a silken cocoon for the pupa. After sufficient development, adults emerge and survive a couple of weeks, during which time they feed on nectar and find a mate. The female then lays small clutches of eggs on vegetation or other objects above the water. About a week later, the newly hatched larvae drop into the water to find a sponge and begin the cycle anew. If the cycle occurs late in the season, spongillaflies overwinter as larvae or as prepupae in cocoons.

Aquatic caterpillars typically feature one to two generations per year. Most larvae pass through five larval instars and reach sizes of 3–35 mm before pupating in a cocoon present in the normal microhabitat occupied by the larva. The adults emerge, pass through relatively short lives spent near the aquatic habitat, and lay single eggs or clusters just below the water surface on aquatic plants or underwater rocks. The selected oviposition site reflects the preferred food source of the future larvae.

III. ECOLOGY, BEHAVIOR, AND ENVIRONMENTAL BIOLOGY

A. Habitats and Environmental Limits

Megalopteran larvae inhabit a moderate diversity of habitats but are most commonly found in cool streams in either riffle areas (hellgrammites and fishflies) or pools with soft sediments (alderflies). Vegetated lake margins are also colonized by these insects, especially by alderflies.

Neuropteran larvae are confined to viable sponge colonies, but these may be in low-sediment streams or shallow waters of lakes. Some individuals colonize exterior crevices of the sponge colony, while others live within the sponge canal system. In the spring in California's Clear Lake, the docks and pilings may be covered with spongillafly pupal cocoons above the water line.

Lepidopteran larvae are most commonly found on the underside of stems of attached or floating aquatic plants in wetlands, lakes, and sluggish lateral areas of rivers or in cases attached to rocks in swift streams and rivers. These living habitats afford a steady supply of both attached algae and living vascular plant tissue as well as the material for protective cases inhabited by some aquatic caterpillars. A few species choose a much different lifestyle by living in the crevices or underside of rocks where they live in silken cases and feed on diatoms and other periphyton.

B. Functional Roles in the Ecosystem and Biotic Interactions

All megalopterans are primarily predaceous on other aquatic invertebrates, including each other, but some species of alderflies will supplement this diet with detritus, especially in winter. Because of their large body size, they can exert a substantial effect on populations of other insects. Megalopterans are in turn preyed upon by each other, by an occasional large invertebrate from another phylum or insect order, and by fish. Indeed, many anglers know the value of capturing these insects for use as bait, and one can occasionally buy live hellgrammites in bait stores near rivers.

All spongillaflies are predators on living sponge colonies. They consume but seem not to kill the colony. In addition to consuming sponge tissue, spongillaflies are known to feed on bryozoans (moss animals) and algae; both additional sources of nutrition often grow near sponge colonies. Because few sponges have predators, the protection afforded to the neuropterans by sponge colony considerably reduces the risk from predation from comparable sized invertebrates or larger species. Competition is also probably not a serious problem for larvae because only a limited number of other invertebrates are obligate or even facultative residents of sponges.

Aquatic caterpillars are all herbivores, but they obtain their nutrition by scraping the surface of plants in search of algae, gouging the outer plant tissue, or burrowing into the plant tissue (a process called boring or mining). They may feed on the damp emergent part of an aquatic macrophyte or graze submerged algae or plant tissue. Some species eat plant seeds. Occasionally, these lepidopterans become sufficiently abundant to impact densities or health of aquatic plant populations.

These caterpillars are somewhat protected by the silken cases but are still subject to predation from other larval insects. In addition, terrestrial parasitoid wasps will implant eggs in any surface crawling caterpillar or pupa they locate, leading to the eventual death of the lepidopteran.

IV. COLLECTION AND CULTURING

Hellgrammites and fishflies are relatively easy to collect from cool, rocky streams using a kick net to stir up the substrate or by turning over large rocks. By contrast, alderflies prefer depositional areas of streams and lakes, where they can be collected with a D-net. Larvae can be raised in an aquarium if it is well oxygenated and sufficient prey items are present. However, these species tend to be cannibalistic, so raising multiple individuals together may be a challenge.

Spongillaflies (Neuroptera) can be collected in streams most easily by collecting sponges, placing them on a pan without water, and waiting for the larvae to crawl out of the sponge as it dries. Maintaining them in the laboratory is difficult for even experts because they require a thriving colony of freshwater sponge for the larvae (which is difficult in itself) and a suitable terrestrial environment for pupation.

Aquatic caterpillars are difficult to spot in the field, but are best found by detaching leaves of floating aquatic plants, placing them in a shallow, well-lighted pan, and searching over the surfaces with a magnifying glass. They can be maintained for their brief lives in an aquarium with living aquatic plants but will eventually pupate and likely emerge as a moth.

V. REPRESENTATIVE TAXA OF HELLGRAMMITES, SPONGILLAFLIES, AND AQUATIC MOTHS

A. Megaloptera

Hellgrammites (Fig. 24a): *Corydalidae*

Identification: Large, cylindrical, worm-like insects, with long lateral filaments on the abdomen and large, heavy jaws. Lacking a single, long tail filament. Head and thorax heavily chitinized. **Size:** 15–90 mm. **Range:** Widespread in temperate and subtropical North America. **Habitat:** Permanent or temporary rivers and streams. In loose sand on rocky bottoms. **Remarks:** These insects are predators and can bite, but it is rarely painful to humans. Pictured here is a larval fishfly.

Hellgrammites (Fig. 24b): *Corydalidae*

Additional Remarks: Pictured here is a larval dobsonfly.

Alderfly Hellgrammites (Fig. 24c): Sialidae

Identification: Cylindrical, worm-like insects, with long lateral filaments on the abdomen and slender jaws. Abdomen with a single, long tail filament. **Size:** 8–14 mm. **Range:** Widespread in subarctic, temperate, and subtropical North America. **Habitat:** Permanent or temporary rivers and streams, burrowing in debris. **Remarks:** Indiscriminate predators and detritivores. Adults do not feed.

B. Spongillaflies (Fig. 24d)

Neuroptera: Sisyridae

Identification: Cylindrical, pudgy insects, with a reduced head and long, slender mouthparts. Slow moving. Typically the color of the sponge host that it is feeding on. **Size:** 3–9 mm. **Range:** Widespread in subarctic, temperate, and subtropical North America. **Habitat:** Lakes, rivers, and streams on freshwater sponges. **Remarks:** Feeds only on freshwater sponges (see Chapter 5).

C. Aquatic Moths (Fig. 24e)

Lepidoptera: Pyralidae

Identification: Cylindrical or flattened caterpillars. The head is typically as wide as or wider than the body. First three leg pairs made up of separate segments and pointed at the tip, remaining legs fleshy, with a ring of small hooks. Some species have many gill filaments per body segment. **Size:** 3–11 mm. **Range:** Widespread in subarctic, temperate, and subtropical North America. **Habitat:** Lakes, rivers, and streams on plants or on rocks. **Remarks:** Moths and butterflies are closely related to the caddisflies (see Chapter 25).

Caddisflies: Insect Order Trichoptera

I. INTRODUCTION TO TRICHOPTERA

Caddisflies are an ecologically diverse and important group of insects, all of which have holometabolous development with aquatic or amphibious larvae. Nearly 1400 species of adult caddisflies have been described in North America, but half the species are known only from their adult stage. Consequently, taxonomic keys to larvae are not completely reliable in all caddisfly families. The large species richness in Trichoptera exceeds that in all other insect orders whose members are primarily aquatic. Most species are confined to streams, and many are familiar to trout anglers (where adults are sometimes called sedges or shadflies). However, about half the trichopteran families contain some species that colonize wetlands, ponds, and lakes. Most caddisflies eat algae or dead organic matter, but a few are almost exclusively predaceous and others depend on an occasional animal prey to boost growth and ultimately reproductive success.

Among insect orders, the closely related Lepidoptera (moths and butterflies) and Trichoptera are best known for their production and use of silk. Caddisflies produce silk from labial glands in the head to construct larval retreats, food-gathering nets, mobile cases for larvae of many species (Fig. 25a), and cocoons for all pupae. The shape and uses of the silken structures vary among caddisflies at the levels of superfamily to species. A few groups are considered "free-living"; that is, they move about underwater without larval cases or hard retreats; these tend to be predaceous species. However, most either build a fixed retreat—as in the retreats of net-spinning, hydropsychid caddisflies—or various shaped cases. These cases, which are found in about half the known species, can often be used for identification of taxa, sometimes even at the generic level, so you should always keep the case with the caddisfly when moving or preserving them. The retreats may be adjacent or attached to a net of varying mesh sizes which is used to capture living organisms (microinvertebrates or algae) or dead organic matter passing downstream in the water column. Other fixed retreats merely serve as shelter during periods when the insect is not foraging for food. Cases come in a large variety of shapes and materials which can characterize a given family. Some cases are shaped like barrels, purses, or saddles and possess no ornamentation or only small amounts of sand, stones, or dead organic material. Others are completely covered by plant material, sand, or stones and may resemble sticks, flattened snails (e.g., the caddisfly *Helicopsyche*), or other objects. This outer material camouflages the animal and also prevents abrasion from sand or minute gravel being transported by the current. Relatively large "rocks" attached to the case can provide some ballast to help prevent the larvae from being displaced downstream by the current. Most cases are long and thin and either round or square in cross section, but some are rather stubby (e.g., cases of *Lepidostoma*). The pupal cases of larvae with cases generally resemble those possessed by the larval stages but are sealed at the ends and attached to the substrate.

Caddisflies are rarely even a minor nuisance to humans, though occasionally loose hairs from the wings of large swarms of adults may irritate the eyes of people or cause an allergic reaction.

ISBN 978-0-12-381426-5, DOI: 10.1016/B978-0-12-381426-5.00025-9

As members of the EPT (Ephemeroptera, Plecoptera, and Trichoptera) monitoring index, caddisflies can be used to evaluate environmental condition and degradation. While most species are much more abundant in clean water, some taxa seem to thrive in somewhat polluted waters.

For an interesting look at the architecture of the larval cases of caddisflies and perspectives on trichopteran classification, see Wiggins (2004).

FIG. 25A. The caddisfly *Oxyethira*, family Hydroptilidae. Note the mobile case of this microcaddisfly.

II. FORM AND FUNCTION

A. Anatomy and Physiology

Caddisfly larvae have elongate, slender bodies that range in length from 2 to 43 mm. They are characterized by very short antennae (visible only with a strong microscope) in most species, a pair of simple eyes, chewing mouthparts, a single tarsal claw on the walking legs, 10 mostly membranous abdominal segments, and a pair of prominent fleshy, nonsegmented prolegs on the last abdominal segment (each proleg has a claw). Rudimentary wings (=wing pads) are absent in larval caddisflies. In some species the legs are fringed with setae that improve either swimming (e.g., in the family Leptoceridae) or extraction of food from the currents (e.g., in Brachycentridae). The head and some thoracic segments can be entirely or partially sclerotized with a thin, horny chitinous patches. The shape and location of this material can be important in classification keys to larvae.

Although most larvae respire by diffusion across the body surface, other species supplement this with oxygen obtained through single or branched abdominal gills. Some larvae of case-building species undulate their body to increase the flow of oxygenated water through the case.

Adults reach body lengths of 2–25 mm and have very long filiform antennae and wings that are profusely covered by short hairs and are folded roof-like over the abdomen when at rest.

B. Reproduction and Life History

Caddisflies are characterized by holometabolous development (egg, larva, pupa, and adult stages). Most species produce only a single generation per year. In warmer climates, however, two generations may emerge each year; while at either high elevations or latitudes, development to adult may require 2 or more years in some species. Adults live for about a month on nectar absorbed by sponge-like mouthparts before laying masses of eggs underwater, on the surface, or just above it on emergent vegetation. Once the larvae hatch after a few week to many months (for diapause eggs in some species), they generally pass through five aquatic larval instars (two to four times as many in some species). They then pupate for several days up to 3 weeks in a sealed larval case or a new silk cocoon. The cocoon is cemented to a large rock or other relatively stable substrate in an area of lower turbulence. The pupal case of some taxa is perforated to allow oxygenated water to reach the pupa; whereas in other species, oxygen simply diffuses through the cocoon walls. The pre-adult cuts its way out of the cocoon and swims to the surface or more likely to the shore where it transforms into a newly molted, winged adult. Caddisflies emerging as adults over water often use the larval skin as a raft on which the adult briefly floats before flying to shore to avoid becoming a hungry fish's meal.

III. ECOLOGY, BEHAVIOR, AND ENVIRONMENTAL BIOLOGY

A. Habitats and Environmental Limits

Trichopterans colonize a broad diversity of habitats from wetlands and lakes to mountain streams and great rivers in all continents except Antarctica. Nearly half the families have species living in lentic ecosystems, but these are almost entirely confined to shallow littoral zones. In rare instances, caddisfly taxa thrive throughout their life cycle in moist terrestrial habitats. And in even more exceptional circumstances, some species colonize marine intertidal and estuarine environments. However, individual species generally have narrow habitat requirements. They are most commonly found in shoreline areas among aquatic vegetation or among rocks in riffle zones. In streams with mobile substrates, the greatest diversity and abundance of caddisflies are often on wood snags that are continually submerged in the stream. This provides a substrate for growth of algal food, a place to erect a food-gathering net in some species, protection from moving substrate, and a potential refuge from insectivorous fish.

Most caddisflies only colonize permanent aquatic habitats. Those that tolerate ephemeral ecosystems typically survive drought conditions as resistant eggs or long-lived adults.

B. Functional Roles in the Ecosystem and Biotic Interactions

Caddisflies occupy all consumer functional feeding groups up through the level of insectivore, and it is not unusual for a larva's trophic position to change with growth and instar number. The majority of species consume algae at some stage and to some degree, but other species tend to specialize on other food types. Species without larval cases or which build retreats and then forage from there tend to be mostly predaceous; these are especially common in the more primitive families Hydrobiosidae and Rhyacophilidae. By contrast, most species that must lug around a case, especially if it is ballasted with rocks or heavy twigs, eat algae either by scraping periphyton from surfaces or by piercing larger filamentous algae to gain a softer, more liquid meal. Heavily ballasted species, like *Helicopsyche*, tend to stay in a limited home range while feeding, but they can reach densities high enough to consume a majority of the available algal production in a rocky stream. In fact, caddisflies in general can effectively suppress benthic periphyton production in many habitats. Other caddisflies with light cases may move long distances (up to 10 m) in a single day in search of food. Competition among larval caddisflies for food has been demonstrated in algal grazers at least during periods between stream spates when the stream bed can be severely disturbed. Caddisflies have also been shown to compete with algal grazers from other insect orders and even different phyla such as snails in the phylum Mollusca.

In headwater streams where deciduous trees dominate the riparian zone, it is not unusual to find trichopterans that specialize in shredding terrestrial leaves which have fallen into the stream. In this case, however, much of the ingested organic matter that is actually assimilated consists of bacteria and other microbes.

Omnivory is a common feeding strategy in many caddisflies. This is especially evident in net-spinning caddisflies. Their silken nets capture dead and living organic matter. When food is scarce, a caddisfly will eat about any organic matter (living or dead) it encounters on the net, attached to its retreat, or living close to the net. However, when food is relatively abundant, it will pick off the dead organic matter from the net to increase filtering and capturing efficiency for live animal prey. Animal prey greatly enhances growth rates and ultimate reproductive success. The net mesh size differs sometimes dramatically among species and thereby influences the volume of water filtered per unit time. Consequently, different meshes vary in capture efficiency and suitability for different flow conditions. For example, net-spinners that specialize on suspended algae have a much smaller mean mesh size than those seeking more animal prey.

Caddisflies in turn fall prey to a variety of vertebrate and invertebrate predators. The latter include stoneflies, hellgrammites, odonates, and a few other trichopterans. They are especially susceptible to bottom-dwelling fish like sculpins and darters.

IV. COLLECTION AND CULTURING

Most larval caddisflies can be collected with an aquatic net either swept through vegetation or used on rocky stream bottoms to capture animals drifting away from the substrate you intentionally disturb. To find net-spinning caddisflies, look on rock surfaces either for the nets directly or for bubbly microturbulence created by water flowing over and through the larval retreats. The entire net can be removed, or the insect can be gently poked out of its retreat. Pupae can be found by examining solid substrates for the attached pupal cases. Place the collected inorganic and organic material in a pan and carefully examine the debris for hidden caddisflies, which may look very much like a small pebble, stick, or piece of moss. Species living in sand or silt are best found by sieving the substrate. Remember to retain the case, where present, with the larvae to improve your ability to identify the caddisfly family.

Caddisflies will survive in a fishless aquarium if provided sufficient oxygen and the correct food, such as microalgae (not the long filamentous forms) growing on rocks for most case-building caddisflies. Predaceous species will require insect prey, such as midge larvae. Maintaining lentic species or those found in stream pools is generally easier than raising caddisflies that normally inhabit turbulent parts of streams where the water is generally cooler and more oxygenated.

V. REPRESENTATIVE TAXA OF CADDISFLIES: ORDER TRICHOPTERA

A. Turtle Caddisflies (Fig. 25b): Glossosomatidae

Identification: Small, slow-moving caddisflies with a heavy stone case that is open at both ends, with both openings ventral. **Size:** 3–10 mm. **Range:** Widespread in North America. **Habitat:** Rivers and streams, usually in strong currents, where they move over rocks scraping up algae and diatoms. May be present on wave-swept lake shores. **Remarks:** These insects may occur in very large numbers in rocky streams. These are the larvae of the little black caddis of anglers.

B. Net Making Caddisflies (Fig. 25c): Hydropsychidae

Identification: Worm-like caddisflies lacking a case and typically holding their body in a "C" shape. Head and thorax well sclerotized. Large obvious gills under the abdomen. **Size:** 3–30 mm. **Range:** Widespread in North America. **Habitat:** Rivers and streams, in strong currents and riffles. May be present on wave-swept lake shores. **Remarks:** These insects build nets on and between rocks to capture particles in the current.

Net Making Caddisflies (Fig. 25d): Philopotamidae

Identification: Worm-like caddisflies lacking a case and not typically holding their body in a "C" shape. Head and thorax well sclerotized. No obvious gills. **Size:** 3–20 mm. **Range:** Widespread in North America. **Habitat:** Rivers and streams, in strong currents and riffles. May be present on wave-swept lake shores. **Remarks:** These insects build nets on and between rocks to capture particles in the current.

C. Freeliving Caddisflies (Fig. 25e): Rhyacophilidae, Hydrobiosidae

Identification: Worm-like caddisflies lacking a case and not typically holding their body in a "C" shape. Gills usually absent, rarely with gills laterally on the abdomen. **Size:** 3–35 mm. **Range:** Widespread in North America. **Habitat:** Rivers and streams, in strong currents and riffles. **Remarks:** These insects are highly mobile predators, and are the larvae of the green sedges.

D. Fixed Retreat Caddisflies (Fig. 25f): Polycentropidae, Dipseudopsidae, Xiphocentronidae, Ecnomidae, Psychomyiidae

Identification: Worm-like caddisflies living in a long silken tube, attached to the substrate. Not typically holding their body in a "C" shape. Head and thorax well sclerotized. Gills usually absent, rarely with gills laterally on the abdomen. **Size:** 3–30 mm. **Range:** Widespread in North America. **Habitat:** Rivers and streams, in strong currents and riffles. **Remarks:** These insects build long, fragile, tubes that may be branched or funnel shaped.

E. Microcaddisflies (Fig. 25g): Hydroptilidae

Identification: Small caddisflies, typically with flattened bodies, often living in a flattened, purse-shaped case made of silk or algae, or building a flattened, fixed, silk retreat. Head and thorax well sclerotized. Gills usually absent, rarely with anal gills. **Size:** 2–6 mm. **Range:** Widespread in North America. **Habitat:** Rivers and streams, in strong currents and riffles. **Remarks:** One rare genus builds its case from pieces of liverworts.

Microcaddisflies (Fig. 25h): Hydroptilidae

Additional Remarks: An example of a flattened fixed retreat.

F. Snailshell Caddisflies (Fig. 25i): Helicopsychidae

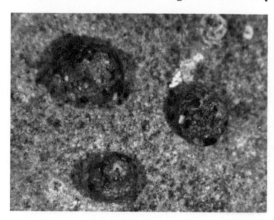

Identification: Small caddisflies in a spiral, snail shell shaped case of sand. **Size:** 2–7 mm. **Range:** Widespread in North America. **Habitat:** Rivers and streams, in moderate currents and riffles. **Remarks:** This is the larva of the trout angler's "Speckled Peters."

G. Typical Caddisflies (Fig. 25j): Limnephilidae, Apataniidae, Beraeidae, Uenoidae, Odontoceridae, Sericostomatidae, Molannidae, Rossianidae, Leptoceridae, Phryganeidae, Lepidostomatidae, Goeridae, Calamoceratidae, Brachycentridae

Identification: Large to medium caddisflies with cases made from sand, gravel, pine needles, leaf fragments, snail shells, or even a hollow twig. **Size:** 2–40 mm. **Range:** Widespread in North America. **Habitat:** Rivers, streams, ponds, lakes, springs. **Remarks:** This figure presents an example of a sand grain case.

Typical Caddisflies (Fig. 25k):

Additional Remarks: An example of a rough plant fragment case.

Typical Caddisflies (Fig. 25l):

Additional Remarks: An example of a smooth plant fragment case.

Typical Caddisflies (Fig. 25m):

Additional Remarks: An example of a spiral plant fragment case.

Typical Caddisflies (Fig. 25n):

Additional Remarks: An example of a square plant fragment case. This is a larva of the Apple Caddis and Grannoms.

Typical Caddisflies (Fig. 25o):

Additional Remarks: An example of a predatory caddisfly: note the long legs used for prey capture. Some predatory species can use the long legs to swim awkwardly. These are the larvae of the Little Brown-Green Sedges.

Typical Caddisflies (Fig. 25p):

Additional Remarks: An example of a sand grain and gravel case.

Chapter 26

Beetles: Insect Order Coleoptera

I. INTRODUCTION TO THE ORDER COLEOPTERA

Although Coleoptera is a huge insect order, only a bit over 3% of the known species have an aquatic phase. Of these, about 1450 species of aquatic beetles have been described from North America, with about half being in predaceous diving beetle family (Dytiscidae). Both adults and larvae are predominately aquatic in about half the families, with the remainder being aquatic primarily only in one life stage. Larvae are a challenge to identify even at the generic level for most groups. By contrast, adult beetles are relatively easy to identify at the generic and even species levels.

Aquatic beetles thrive in almost all aquatic habitats, but they are most common in lentic ecosystems where they may frequent weedy areas or on the water surface. Lotic species are more common away from the main currents, but riffle beetles (Elmidae) tolerate moderately fast waters. Beetles on the whole are not unusually sensitive to water pollution compared to many other aquatic species. Some, however, are found only in clean, well-oxygenated habitats.

Aquatic beetles rarely pose a threat to humans, unless you mishandle a predaceous diving beetle or similar adult predatory species.

II. FORM AND FUNCTION

A. Anatomy and Physiology

Larvae are highly diverse in structure, even more so than adults. Most are long and somewhat cylindrical in shape, but water pennies (Psephenidae) are flattened and round, which helps them resist being dislodged by stream currents. Beetle larvae can be distinguished from other insects by a combination of a heavily sclerotized head capsule, prominent mandibles and other mouthparts, and antennae with usually two to three segments (except in the family Scirtidae, in which the antennae are long and whip like). The thorax and abdomen are also hardened in many species. External gills are present on the abdomen of some species, but in most cases these are not prominent. No prolegs, long terminal filaments, or wing pads occur in larval beetles. Larvae are highly variable among species in length, ranging from 2 to 70 mm.

Adult aquatic beetles have a wide diversity of body sizes, ranging from minute beetles less than 1 mm long to predaceous diving beetles of 40 mm. They are in general quite variable in structure, all having chewing mouthparts, antennae of 11 or fewer segments, compact and heavily sclerotized bodies, and shell-like fore wings (elytra) extending from the middle thoracic area and covering the membranous hind wings and most of the thorax and abdomen.

ISBN 978-0-12-381426-5, DOI: 10.1016/B978-0-12-381426-5.00026-0

Larval beetles obtain oxygen mostly by diffusion across the skin surface, but adults have a chitinous shell that greatly hinders integumentary respiration. Instead, they have spiracles (internal breathing tubes with an external opening) just like fully terrestrial beetles and other insects. The trick is to absorb the air without swamping the spiracles. Many species have hairs surrounding the spiracles and certain other surfaces which repel water (hydrofuge or hydrophobic hairs). To get air to the spiracles, adult beetles may stay constantly at the surface, make occasional trips there, or carry a bubble under water as a physical "gill." The bubble may be periodically renewed from the surface, or the beetle relies on oxygen diffusing into the bubble and carbon dioxide diffusing out; the latter works especially well in colder environments where water carries more oxygen. Some species obtain oxygen under water by piercing the vascular tissue of aquatic weeds to obtain bubbles of oxygen passing through the tissues.

Crawling is the typical mode of locomotion for larvae and most adults. However, some adults have long and profuse hairs on at least one pair of legs to aid swimming. In some other species, such as whirligig beetles (family Gyrinidae), one or more pairs of legs are greatly flattened to act as paddles.

An interesting adaptation to life at the water surface occurs in whirligig beetles. Their compound eyes are divided in halves, with the upper portion adapted to seeing predators or prey above water and the lower portion suitable for viewing similar objects underwater.

B. Reproduction and Life History

Life history characteristics of aquatic beetles vary greatly, in part because an aquatic existence evolved repeatedly in many distinct groups of terrestrial beetles. Consequently, the larval stage is the longest in some taxa while the adult stage is more extensive in other groups. The most common pattern is for an aquatic larva to pass through three to eight instars over an average of 6–8 months before pupating on land for 2–3 weeks and then emerging as an adult which mates and then lays eggs singly or in masses underwater on a hard surface or in a damp area near the stream or lake. Adults living in terrestrial habitats tend to survive only several weeks, while those inhabiting aquatic systems as adults can live for months or even longer and not reproduce until the spring. Univoltine life cycles are the most common, but some species reproduce twice during the year and others require 2 or more years to complete their life cycles. One of the longer living coleopterans is the wood-boring aquatic beetle (*Lara avara*, Elmidae) which is known to live upwards to 5 years. This long life cycle probably reflects the low nutritional quality of the wood it regularly eats which slows maturation. In general, large beetles, such as predaceous diving beetles and water scavenger beetles (Hydrophilidae), tend to live the longest.

III. ECOLOGY, BEHAVIOR, AND ENVIRONMENTAL BIOLOGY

A. Habitats and Environmental Limits

Unlike some other insect orders described in this field guide, aquatic beetles are most abundant in wetlands, ponds, and lakes. While many species occur in streams, they tend to live in areas of slow to zero currents. Some exceptions include many riffle beetles found in moderate currents, where their populations can constitute a sizeable portion of the animal community biomass. Beetles are one of the more common subterranean insect species, and a few taxa colonize geothermal springs. Most of the latter cannot survive temperatures much above 30°C, but a couple of species can live at 45°C. At the opposite end of the thermal scale, some beetles thrive on ice fields.

Adult beetles can disperse to more favorable habitats if conditions in their normal ecosystem become seasonally unfavorable or if food resources are too low. In northern lakes subject to complete freezing, adult beetles will often seek lotic habitats to overwinter in areas where the stream does not freeze solid to the bottom. In ephemeral ponds, beetles will either survive the period of desiccation in a resistant egg stage or fly to another pond with water.

Although the vast majority of aquatic beetles colonize freshwater habitats, a few taxa live in estuarine and marine habitats, including the demanding marine intertidal zone.

B. Functional Roles in the Ecosystem and Biotic Interactions

Most water beetles are predators as both larvae and adults, but other taxa are collector-gatherers or they feed or algae (e.g., many riffle beetles). A few species can even thrive on a diet of cyanobacteria, which is toxic or at least distasteful to most other herbivores. The predators can be divided into engulfers and piercers. The latter secrete fluids into the prey and then consume the liquified tissue. Other invertebrates are the most common prey item, but late instar predaceous diving beetles (also called water tiger beetles) can capture small fish and amphibians.

The relatively large size of many aquatic beetles as larvae and adults and the habit of some species for living at the surface make them potential prey for fish and birds. Most species attempt to hide from predators, but other beetles rely on their hard and spiny bodies to protect them or they are fast or produce distasteful or irritating chemicals from repugnatory glands.

IV. COLLECTION AND CULTURING

Many adult beetles can be collected with an aquatic net from shore or by wading in shallow water. Sweep the net through aquatic weeds in search of specimens. Riffle beetles can be collected from rocks, wood snags, or trailing vegetation in some streams. Bottle traps are sometimes needed for larger species. Surface-swimming species like whirligig beetles can be herded toward shore to improve your capture efficiency.

Larval beetles are relatively easy to maintain in an aquarium although many are too small to be readily visible. Those requiring flowing water are clearly the hardest to rear. Adults in some families, such as Elmidae and Dyticidae, can be kept in aquaria for a year or more.

V. REPRESENTATIVE TAXA OF AQUATIC BEETLES: ORDER COLEOPTERA

A. Skiff Beetles (Fig. 26a): Hydroscaphidae

Identification: Small, black or brown beetles with the abdomen projecting well past the wing covers. Larvae (not shown) are flattened anteriorly and tapering posteriorly, with the head broad. **Size:** 1–2 mm. **Range:** Arizona and southern California and Nevada. **Habitat:** On rocks covered in algae in shallow stream margins. **Remarks:** These insects may occur in very large numbers in rocky streams.

B. Round Sand Beetles (Fig. 26b): Carabidae

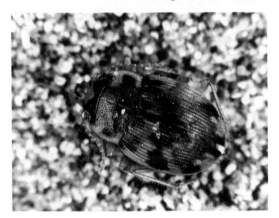

Identification: Small, disc-shaped, convex beetles typically yellow with green or dark markings. **Size:** 5–8 mm. **Range:** Widespread from southern Canada, south. **Habitat:** On or in damp, loose sand on river and stream margins. **Remarks:** To collect these animals, pour water on damp sand. If they are present, they will come to the surface to escape drowning.

C. Whirligig Beetles (Fig. 26c): Gyrinidae

Identification: Small- to medium-sized, oval-shaped, convex, black beetles. Front legs are long and normally shaped, whereas the middle and hind legs are modified as short broad paddles. The larvae are long bodied, with long, unbranched lateral gills, and a long rectangular head. **Size:** 3–16 mm. **Range:** Widespread in North America, with the larger species in the east. **Habitat:** Actively and rapidly swimming on the surface of lakes, ponds, and slower portions of rivers and streams. Usually swimming in circular or "figure eight" patterns. **Remarks:** Sometimes occurs in large numbers.

D. Water Crawling Beetles (Fig. 26d): Haliplidae

Identification: Small, thick-bodied, spindle-shaped beetles, typically yellow with green or dark markings. The abdomen is covered ventrally by the broadly expanded hind leg plates. **Size:** 1.7–4.5 mm. **Range:** Widespread. **Habitat:** In vegetation in ponds, lakes, and seasonal wetlands such as vernal pools. On or in damp, loose sand on river and stream margins. **Remarks:** Most species are poor swimmers; however, one western flightless species swims very well.

Water Crawling Beetles (Fig. 26e): Haliplidae

Additional Remarks: Example of a typical Water Crawling Beetle larva. Note the long, filamentous, dorsal spines. Some species lack these spines.

E. Trout Stream Beetles (Fig. 26f): Amphizoidae

Identification: Large black beetles with a narrow head. Legs lacking any swimming hairs, and they swim very poorly. **Size:** 11–16 mm. **Range:** Western mountains. **Habitat:** Occurs in log and debris jams at the water surface in cool to cold snowmelt streams and rivers. **Remarks:** These beetles have defensive scent glands at the tail that produce a thick liquid smelling like rotting wood.

Trout Stream Beetles (Fig. 26g): Amphizoidae

Additional Remarks: Example of the larva. Note the broad plates extending from the sides of the animal.

F. Burrowing Water Beetles (Fig. 26h): Noteridae

Identification: Small black or brown extremely streamlined beetles. Legs with swimming hairs, and the base of each hind leg with a broad angular keel. Fast swimmers. **Size:** 1–5 mm. **Range:** Rare in the USA and eastern Canada. More common in the southeastern USA. **Habitat:** Occurs in aquatic plants in pools, ponds, lakes, and slow rivers and streams. **Remarks:** Very easily confused with the Predaceous Diving Beetles. A microscope is typically needed to tell small members of the two families apart.

G. Predaceous Diving Beetles (Fig. 26i): Dytiscidae

Identification: Highly variable with more than 4000 species worldwide, fast swimming aquatic beetles (many examples below). Most species are very small and black or brown. Larger species may be more spectacular in coloration. **Size:** 1.6–45 mm. **Range:** Very common and widespread around the world. **Habitat:** Occurs in all aquatic habitats, except the swiftest of waters. **Remarks:** Small specimens are very easily confused with the Burrowing Water Beetles.

Predaceous Diving Beetles (Fig. 26j): Dytiscidae

Additional Remarks: An example of a larval Predaceous Diving Beetle. The larvae of the larger species are known as water tigers (see also Fig. 4h).

Predaceous Diving Beetles (Fig. 26k): Dytiscidae

Additional Remarks: An example of a swift water species.

Predaceous Diving Beetles (Fig. 26l): *Dytiscidae*

Additional Remarks: An example of a Predaceous Diving Beetle.

Predaceous Diving Beetles (Fig. 26m): *Dytiscidae*

Additional Remarks: An example of a Predaceous Diving Beetle. The Gold Spotted Diving Beetle is common in the southwest deserts.

Predaceous Diving Beetles (Fig. 26n): *Dytiscidae*

Additional Remarks: An example of a Predaceous Diving Beetle.

Predaceous Diving Beetles (Fig. 26o): Dytiscidae

Additional Remarks: An example of a Predaceous Diving Beetle.

Predaceous Diving Beetles (Fig. 26p): Dytiscidae

Additional Remarks: An example of a Predaceous Diving Beetle.

Predaceous Diving Beetles (Fig. 26q): Dytiscidae

Additional Remarks: An example of a Predaceous Diving Beetle. The genus *Colymbetes* is the only beetle genus in the world to have transverse grooves in the wing covers.

Predaceous Diving Beetles (Fig. 26r): Dytiscidae

Additional Remarks: An example of a Predaceous Diving Beetle.

H. Water Scavenger Beetles (Fig. 26s): Hydrophilidae

Identification: Highly variable, with nearly 2500 species worldwide. Some species are good swimmers, but most swim poorly (many examples below). Most species are very small and green, black, or brown. **Size:** 2–40 mm. **Range:** Very common and widespread around the world. **Habitat:** Occurs in all aquatic habitats, except the swiftest of waters. Some occur on shore margin habitats. Often in vegetation, or under rocks. **Remarks:** The species depicted here is a Minute Mud Loving Beetle, which lives on sand or mud at the shore line, and is often coated in mud.

Water Scavenger Beetles (Fig. 26t): Hydrophilidae

Additional Remarks: An example of a *Helophorus* beetle, which lives on emergent vegetation.

Water Scavenger Beetles (Fig. 26u): Hydrophilidae

Additional Remarks: An example of an *Epimetropus* beetle.

Water Scavenger Beetles (Fig. 26v): Hydrophilidae

Additional Remarks: An example of an *Hydrochus* beetle, another Water Scavenger Beetle that lives at the surface and margins of aquatic habitats.

Water Scavenger Beetles (Fig. 26w): Hydrophilidae

Additional Remarks: An example of an sculptured Water Scavenger.

Water Scavenger Beetles (Fig. 26x): Hydrophilidae

Additional Remarks: An example of a typical Water Scavenger Beetle larva.

Water Scavenger Beetles (Fig. 26y): Hydrophilidae

Additional Remarks: An example of a hump-backed water beetle.

Water Scavenger Beetles (Fig. 26z): Hydrophilidae

Additional Remarks: An example of a humped back water beetle larva.

Water Scavenger Beetles (Fig. 26aa): *Hydrophilidae*

Additional Remarks: An example of a typical small Water Scavenger. These insects are usually found in ponds and wetlands in vegetation.

Water Scavenger Beetles (Fig. 26ab): *Hydrophilidae*

Additional Remarks: An example of another typical small water scavenger (*Cymbiodyta*) found in springs and stream vegetation.

Water Scavenger Beetles (Fig. 26ac): *Hydrophilidae*

Additional Remarks: An example of an *Hydrochara* beetle. These water scavengers can be very large and some species are bright green.

Water Scavenger Beetles (Fig. 26ad): Hydrophilidae

Additional Remarks: An example of a *Tropisternus* beetle. These are extremely common medium-sized beetles in ponds, wetlands, and seasonal streams where fish do not occur.

Water Scavenger Beetles (Fig. 26ae): Hydrophilidae

Additional Remarks: Another example of a *Tropisternus*, this one with yellow margins.

I. Minute Moss Beetles (Fig. 26af): Hydraenidae

Identification: Tiny, narrow-bodied beetles that swim poorly. Most species are black or brown. **Size:** 1.2–2.5 mm. **Range:** Widespread. **Habitat:** Occurs in algae, moss, vegetation or clinging to woody debris at or near the margins of wetlands, ponds, lakes, rivers, and streams or springs. **Remarks:** Often mistaken for certain Water Scavenger Beetles.

J. Riffle Beetles (Fig. 26ag): Elmidae

Identification: Small beetles that do not swim. Typically brown or black in color, but may have orange, red, or yellow markings. **Size:** 1–8 mm. **Range:** Very common and widespread around the world. **Habitat:** Found on rocks or woody debris in swift water. **Remarks:** Riffle Beetles are indicators of well-oxygenated water.

Riffle Beetles (Fig. 26ah): Elmidae

Additional Remarks: The genus *Optiocervus* is reported to be unpalatable to fish.

Riffle Beetles (Fig. 26ai): Elmidae

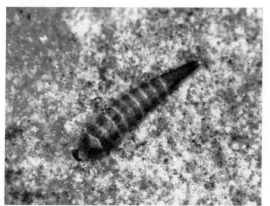

Additional Remarks: Depicted here is a typical larval Riffle Beetle.

Riffle Beetles (Fig. 26aj): *Elmidae*

Additional Remarks: Depicted here are an adult and a larva of the Riffle Beetle *Ancyronyx*. This species is common in the southeast, and is easily recognized by its long, grappling hook like legs.

Riffle Beetles (Fig. 26ak): *Elmidae*

Additional Remarks: The larva of the Riffle Beetle *Lara* is spiny and feeds on decomposing wood.

K. Long Toed Water Beetles (Fig. 26al): Dryopidae

Identification: Medium-sized beetles that are brown, gray, or black, with short thick antennae, long legs, and long claws. **Size:** 4–8 mm. **Range:** Widespread in North America, but only really common in the southwestern deserts. **Habitat:** Found on sand, rocks, or woody debris in swift water. **Remarks:** These beetles feed on vegetation.

L. Ptilodactylid Beetles (Fig. 26am): Ptilodactylidae

Identification: Larval beetles with a long cylindrical body and a tuft of gills at the end of the abdomen. **Size:** 2–16 mm. **Range:** Uncommon. Found in the far western USA and in the eastern portions of Canada and the USA. **Habitat:** Found under rocks or debris in swift water. Many species are found in springs. **Remarks:** The adults are terrestrial, only laying their eggs in the water.

M. Forest Stream Beetles (Fig. 26an): Eulichadidae: *Stenocolus scutellaris*

Identification: Larval beetles with a long cylindrical body and rows of gills under the abdomen. **Size:** 12–18 mm. **Range:** Montane and foothill regions of California. **Habitat:** Found under rocks or debris in swift water. **Remarks:** The adults are terrestrial, only laying their eggs in the water.

N. Varigated Mud Loving Beetles (Fig. 26ao): Heteroceridae

Identification: Flattened, mottled, small beetles, frequently with very long jaws. Front legs and head used for digging. **Size:** 1–8 mm. **Range:** Widespread in the Americas. More common in Canada. **Habitat:** Burrowing in fine sand or mud at the margins of rivers, lakes, streams, and wetlands. **Remarks:** To collect these animals, splash water on damp sand or mud at the water's edge. If they are present they will come to the surface to escape drowning.

O. Water Pennies (Fig. 26ap): Psephenidae

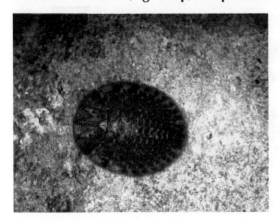

Identification: Flattened, circular, disc-shaped, or oval larval beetles, resembling trilobites. Generally brown to orange. **Size:** 3–8 mm. **Range:** Eastern North America and montane regions and lowlands of the far west. Rare or absent in the southwest deserts. **Habitat:** On rocks or large pieces of wood in swift, well-oxygenated permanent rivers and streams. **Remarks:** The adults are rare, only living for a short period in spring. They may be found on rocks depositing their eggs.

P. Leaf Beetles (Fig. 26aq): Chrysomelidae

Identification: Oblong beetles with metallic markings. Base color may be black, gray, blue, green, or yellow. **Size:** 4–9 mm. **Range:** Widespread in North America. **Habitat:** On emergent vegetation, rarely occurring below the water surface. **Remarks:** There are numerous species, each species feeding only on one species of aquatic plant.

Q. Aquatic Weevils (Fig. 26ar): Curculionidae

Identification: Cylindrical beetles with a projecting snout bearing the mouthparts apically. Brown, gray, black, or mottled. **Size:** 3–7 mm. **Range:** Widespread in North America. **Habitat:** Feeding on aquatic plants, typically emergent plants in lakes and ponds, sometimes in slow rivers. May climb down an emergent plant under water to feed. **Remarks:** Numerous species, with many undescribed species in the West.

Chapter 27

Midges, Mosquitoes, Blackflies, and Other True Flies: Insect Order Diptera

I. INTRODUCTION TO THE ORDER DIPTERA

This order of true flies contains more species with aquatic stages than any other insect group, and most of the species remain undescribed, in part because scientists have not linked most terrestrial adults with their aquatic larvae and pupae. Included among the roughly 5400 described species of Diptera in North America are the following families with at least 200 species each: biting midges (Ceratopogonidae), true midges (Chironomidae), blackflies (Simulidae), craneflies (Tipulidae), shoreflies (Ephydridae), aquatic muscids (Muscidae; their more common terrestrial relatives are called houseflies), and horseflies (Tabandidae). Of the nearly 40% of aquatic insects represented by this order, a third are true midges.

The great species diversity of dipterans is also reflected in their ecological diversity. They include many species of invertebrate predators, herbivores, and detritivores. Dipterans are found in every inland water habitat from ephemeral ditches to the Great Lakes and our largest rivers, and they are also very common in estuarine habitats but do not live in the open ocean. Some are restricted to very clean waters, but others can tolerate highly polluted environments.

Unlike all other invertebrate orders described in this field guide, dipterans contain many species that as adults are harmful or at least annoying to humans. At the top of the list are mosquitoes (family Culicidae with 173 North American species), but other noxious groups include blackflies, biting midges, deer flies, and horseflies. Many adult dipterans transmit parasites or diseases that can be debilitating or fatal to humans, such as malaria and West Nile virus. Other flies emerge in such high numbers (some midges) that they can cause allergic reactions in humans or clog air conditioning units. On the other hand, dipteran larvae are extremely important in aquatic food webs leading up to game fish, and some groups are raised in hatcheries as fish food, and many adult dipterans are important plant pollinators or food for dragonflies and birds such as swallows, swifts, phoebes, and flycatchers.

II. FORM AND FUNCTION

A. Anatomy and Physiology

Dipteran larvae can be distinguished from most other insects by their lack of segmented thoracic legs. Instead of the traditional jointed legs, many groups have one or more pairs of fleshy, locomotory

ISBN 978-0-12-381426-5, DOI: 10.1016/B978-0-12-381426-5.00027-2

prolegs on the thorax and/or abdomen, each with curved or even hook-like spines. Fleshy turbercles occur in some species and serve both sensory and locomotory functions. The head of a larva may be exposed and heavily sclerotized, as in midges, or strongly reduced and only partially projecting (sometimes only with mouthparts protruding). The thorax and abdomen are usually fleshy, sometimes with scattered sclerotized plates, and the entire body is typically tubular and long; average lengths are 2–25 mm but may reach 10 cm in some species. Wing pads are always absent in larvae but present in pupae.

Many larvae respire across the skin, and small gills are present in some taxa. Others dipterans obtain oxygen from the atmosphere using spiracles and either long or short breathing tubes (as in mosquitoes). A few groups extract oxygen from plant tissues. Some true midges that frequent somewhat anoxic habitats, like blood worms (a type of true midge), have an invertebrate form of the respiratory pigment hemoglobin which aids in capturing oxygen molecules.

Adult dipterans range in length from 1 to 12 mm but relative giants of 25–60 mm are known; the latter include large craneflies. They have long tubular bodies and a single pair of membranous wings; the hind wings are rudimentary and nonfunctional for flying. In feeding adults, the mouthparts are adapted for consuming liquid food using either blunt pads for sponging up liquid or sharp tubes for penetrating flesh and sucking up liquids, as in mosquitoes. Even though craneflies are sometimes called "mosquito hawks," they do not eat mosquitoes, nor will they bite humans.

B. Reproduction and Life History

All aquatic flies have holometabolous (complete) development with egg, larval, pupal, and adult stages. All but the adult stage lives in water in most cases. The lifespan of an adult varies by species and may extend from a few hours to weeks or months. Females lay their eggs singly or in clumps, usually in water and sometimes attached to objects. Eggs tend to last only a few days except for diapause eggs which are used to circumvent inhospitable temperatures or the lack of water in the ecosystem. After hatching, the larvae of most species pass through three to four instars (six to seven in blackflies) before pupating near the bottom, at the water surface, or on land. The larval stage lasts anywhere from about 2 weeks to several months.

While a multivoltine life cycle is most common, some dipterans only produce one generation per year (univoltine) and may require multiple years to produce a new generation in arctic and other cold environments. The shorter generation times characterizing this order enable dipterans to respond quickly to new food or other resources and to recover rapidly from unfavorable conditions such as stream spates.

III. ECOLOGY, BEHAVIOR, AND ENVIRONMENTAL BIOLOGY

A. Habitats and Environmental Limits

The diversity of habitats exploited by dipterans is exceeded by no other aquatic insect order. If the habitat exists for at least 2 weeks and adults have access to the surface, a reproducing population of Diptera will be associated with it. This includes all lentic and lotic epigean habitats, saline pools, estuaries, and intertidal marine pools. Many thrive in open water habitats, but others spend their short larval existence in damp mud with no areas of open, standing water. Some also colonize temporary aquatic habits formed in trees (e.g., tree hole mosquitoes), epiphytic plants, pitcher plants, or other aquatic phytotelma habitats. Moreover, dipterans and coleopterans are the most frequently encountered insects in subterranean waters. Dipterans can also be found in such harsh environments as geothermal pools (up to 50°C), alkaline and salt lakes (brine flies living in waters at least up to eight times the ocean's salinity), acidic pools down to a pH of 2.0, and temporarily anoxic ponds with heavy organic pollution. They have colonized all continents including Antarctica.

With the exception of a few planktonic groups such as mosquitoes and phantom midges, most dipteran larvae live in or on bottom substrates or on or within plants. Midges dominate the diversity and biomass of insects at all depths, including the sublittoral and deeper areas of lakes. Late instar phantom midges reside near the bottom during the day to avoid planktivorous fish and then migrate upward at night in search of zooplankton prey.

B. Functional Roles in the Ecosystem and Biotic Interactions

This order includes a very broad range of trophic groups including species with larvae that are herbivores (scrapers and plant piercers), shredders, collector-gatherers, filterers, detritivores, and predators. Sometimes a single family includes species representing each of these functional feeding groups. Most predators seek dipterans and other benthic prey, except for a few planktivorous predators such as phantom midges. Many dipterans, especially herbivores, collector-gatherers, and detritivores, construct somewhat flimsy cases of fine organic or inorganic material glued together with salivary secretions. The cases that camouflage most fly larvae may also serve as a net to collect food in suspension for filter feeding dipterans. In those instances, the dipteran either produces a water current or exploits that already present in the surrounding stream to bring a steady supply of food to the net. Most backfly larvae are also filterers, but they use setose head appendages (labral fans) to filter diatoms and other food from the current; some data suggest that blackflies can extract bacteria and dissolved organic matter from the water as food sources. Other blackfly larvae graze on benthic algae.

Most adult dipterans either do not feed or they consume nectar or other liquid food, such as sap or vertebrate blood. However, some adults also prey on aquatic invertebrates or act as parasitoids and implant their eggs in the tissue. For example, the moderate-sized family Sciomyzidae (175 species in North America), known as marsh flies, have larvae and adults that are predators or parasitoids of snails and fingernail clams.

Predation on dipteran larvae can be intense from other dipteran taxa and insects in other orders as well as from other invertebrates and vertebrates (especially fish). In hydrologically dynamic sand-bed rivers, such as those in the US Great Plains, where copepod and cladoceran zooplankton are rare, dipterans may be the main source of food for larval and young-of-the-year fish. They are certainly the principal pathway in these rivers for food to pass from the algae to higher trophic levels because of their dominance in the benthic community. Adult dipterans are eaten by birds, dragonflies, and some other insects.

IV. COLLECTION AND CULTURING

Aquatic dipterans are best collected by three different means depending on where the dipterans live. Midges and other infaunal species are easiest to collect by scooping up substrate, sieving the silt, sand, or detritus, and then looking for individuals in the retained contents in a white tray. This can also work for many other species which live in leaf packs (e.g., craneflies). Pulling a D-net or a plankton net through aquatic vegetation in a pond, pool, or ditch lacking fish will often reveal larval mosquitoes. And, of course, you can find mosquitoes and a few other taxa throughout the summer in most containers in your backyard that have held water for a few days to weeks. Larval blackflies occur mostly in streams in the upper USA and northward where they can be collected on bedrock in medium to small streams with moderate to high currents. Place any collected organic matter in a white tray with water; the larvae will commonly wiggle free, thereby making them more visible.

Rearing dipterans in a home aquarium is possible; however, the challenge is that these insects have short generation times, meaning that you will soon be faced with adults flying around your house if you are not cautious! Larval chironomid (true) midges, such as those of the large red blood worm *Chironomus* spp., are useful for feeding predaceous invertebrates and fish in your aquarium, and they are relatively easy to raise in the laboratory in a way that prevents most adults from escaping (and the adults do not bite humans anyway). To raise them, first obtain a gallon glass jar (like a large pickle jar) and be ready to cover the top with a mesh material such as cheesecloth, porous nylon, or even a material called bridal veil. Clean the jar with water only to avoid any toxic chemicals. The cloth should be tightly bound to the glass edge with a strong rubber band but "bloused out" to provide some perching areas. The purpose of this mesh is to allow air into the jar while permitting the adults to emerge, have a place to rest and mate, and have access to the water to lay their eggs. Obtain enough paper towels to fill the jar and cut them in strips of about an inch wide (we have used the brown paper towels common in restrooms or research laboratories, but kitchen towels would probably also work). Soak the towels in water from a stream or pond for a day or so, dump out the water, and then grind the towels to a slush in a blender with a couple of tablespoons of fish food from a pet store. Place an aquarium airstone and air hose into the bottom of the gallon jar, with the hose passing through the mesh (fix the mesh tightly around the hose to prevent adults from later escaping). This will prevent conditions from becoming completely anoxic. Return the blender-ground material to the gallon jar, fill the jar with water from the field up to about 2 inches from the top, and allow the mixture to develop a colony of bacteria over a period of about 1–2 weeks. You can then add live chironomid larvae obtained from a supply house or collected in the field. Look for blood worms in low oxygen sediments. We have found them readily in shallow organic pools remaining on sandbars when the river water has receded; they can also be found in organically rich backwater areas where oxygen is low but not entirely absent. You can occasionally add small amounts of fish food to the aquarium (no more than every 2 weeks), but be cautious not to overdo it and cause the water to foul. Populations of midges will continue to renew themselves in the jars, thereby providing a steady source of invertebrate prey for dragonfly nymphs and other predators in your aquarium. As the organic material begins to drop, start a new colony with midges from the old jar.

V. REPRESENTATIVE TAXA OF AQUATIC FLIES: ORDER DIPTERA

A. Craneflies (Fig. 27a): Tipulidae, Tanyderidae

Identification: Variable aquatic maggots, often brown or white, with a ring of tentacles at the tail end. An obvious chitinized head is lacking. **Size:** 3–50 mm. **Range:** Widely distributed. **Habitat:** In sand or gravel of swift streams, or in organic debris in quiet margins of rivers, lakes, streams, ponds, and wetlands. **Remarks:** Numerous species; some are predators, some are herbivores feeding on diatoms and plant tissue, including roots.

Craneflies (Fig. 27b): Tipulidae, Tanyderidae

Additional Remarks: An example of one of the larger cranefly larvae, *Dicranota*.

Craneflies (Fig. 27c): Tipulidae, Tanyderidae

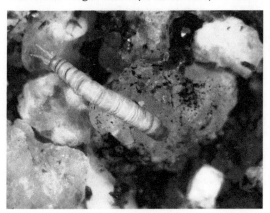

Additional Remarks: An example of one of the smaller cranefly larvae, *Antocha*. This species has enlarged welts for creeping along the substrate.

Phantom Craneflies (Fig. 27d): Ptychopteridae

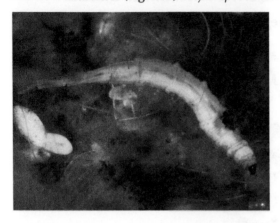

Identification: Elongated white aquatic maggots, with fine, short hairs and an elongated, retractable breathing siphon from the tail. An obvious chitinized head is present, but often retracted. Rings of creeping welts encircle the body. **Size:** 3–8 mm. **Range:** Uncommon. Mostly found in the western USA, and a few species from the eastern USA and Canada. **Habitat:** In heavily vegetated streams, living in saturated mud with a high organic content near the stream margin. **Remarks:** The adults are rare, with some species having modified front legs for catching breezes and carrying the adult long distances.

B. Mothflies (Fig. 27e): Psychodidae

Identification: Elongated white aquatic maggots, with transverse rows of short hairs, or oval to oblong creeping insects with large transverse chitinous plates. An obvious chitinized head is present. Creeping welts may or may not be present. **Size:** 1–4 mm. **Range:** Widespread and common. **Habitat:** Some species occur on rocks, sometimes in riffles. Other species live in organic rich quiet areas of ponds and wetlands. **Remarks:** In addition to their distribution in nature, these animals sometimes occur in shower or sink drains among old hair clogs.

Mothflies (Fig. 27f): Psychodidae

Additional Remarks: An example of the creeping armored form.

C. Midges

Net Winged Midges (Fig. 27g): *Blephariceridae*

Identification: Solid, stocky, cylindrical insects, resembling caterpillars with their several pairs of fleshy prolegs. An obvious chitinized head is present. **Size:** 1–7 mm. **Range:** Widespread and common. **Habitat:** On rocks in swift, well-oxygenated water, including on waterfalls. **Remarks:** Most adult females capture and consume other flies, including mosquitoes. It is unknown whether adult males feed.

Net Winged Midges (Fig. 27h): *Blephariceridae*

Additional Remarks: The pupa resembles a limpet. It can easily be recognized by the segmentation of the body and the breathing organs projecting from the head.

Mountain Midges (Fig. 27i): *Deuterophlebiidae*

Identification: Flattened insects, resembling caterpillars with their several pairs of laterally projecting fleshy prolegs. An obvious chitinized head is present with long forked antennae. **Size:** 3–5.5 mm. **Range:** Western Canada and USA. **Habitat:** On smooth light-colored rocks that have cracks and pittings in very swift, well-oxygenated water in mountain streams, including on waterfalls near the water surface. Never on multicolored or dark-colored rocks, or deeper than cm. **Remarks:** The pupae are similar to that of the net-winged midge described above.

Dixid Midges (Fig. 27j): *Dixidae*

Identification: Elongated aquatic maggots with an obvious chitinized head bent upward. Some body segments with a dorsal ring of dense hairs. Tend to move with the body folded in a "u" shape. **Size:** 3–5 mm. **Range:** Common and widespread. **Habitat:** These animals live at the water surface of ponds and wetlands where fish are absent, or occur at the margins of lakes, ponds, and streams or at the surface in dense vegetation or algae where fish are present. **Remarks:** The pupae are similar to those of mosquitoes, below.

Chironomid Midges, Bloodworms (Fig. 27k): *Chironomidae*

Identification: Elongated, worm-like aquatic maggots with an obvious chitinized head directed forwards or down. Tend to move by thrashing the ends of the body back and forth. Usually one pair of fleshy prologs near the head. **Size:** 1–6 mm. **Range:** Common and widespread. **Habitat:** Midge larvae are found in all aquatic habitats, either swimming or moving in the open water of fishless wetlands, or building tubes on rocks, in sediment, or in plants, or boring into algae. **Remarks:** The pupae are similar to those of mosquitoes, below.

Chironomid Midges, Bloodworms (Fig. 27l): *Chironomidae*

Additional Remarks: An example of a bloodworm midge larva. The red color is due to hemoglobin pigments that help the animal to breathe in low oxygen conditions.

Mosquitoes (Fig. 27m): Culicidae

Identification: Elongated, worm-like aquatic maggots with the head and thorax broadly expanded. The head is chitinized and obvious. Tend to move by thrashing the ends of the body back and forth. Prolegs are absent. Breathing occurs through breathing siphon at the tail. **Size:** 1–8 mm. **Range:** Common and widespread. **Habitat:** Typically found in fishless ponds, wetlands, tree holes, and artificial habitats. Some species live in pockets on the tops of algal mats where fish occur. **Remarks:** The larvae do not bite.

Mosquitoes (Fig. 27n): Culicidae

Additional Remarks: An example of an *Anopheles*. Note the extremely short breathing tube. These animals may occur in algal mats.

Mosquitoes (Fig. 27o): Culicidae

Additional Remarks: An example of a mosquito pupa.

Phantom Midges (Fig. 27p): *Chaoboridae*

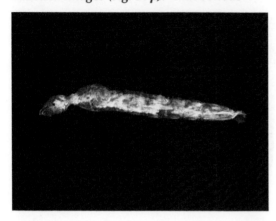

Identification: Elongated, worm-like aquatic maggots with enlarged antennae. The head is chitinized and obvious, sometimes broad, bent downward on an obvious "neck." Prolegs are absent. The breathing siphon at the tail is not at all obvious. **Size:** 1–12 mm. **Range:** Common and widespread. **Habitat:** Pools, ponds, wetlands, lakes. **Remarks:** These insects prey on mosquito larvae and some zooplankton like water fleas (Cladocera).

Phantom Midges (Fig. 27q): *Chaoboridae*

Additional Remarks: An example of a broad headed phantom midge.

Biting Midges, No-See-Ums (Fig. 27r): *Ceratopogonidae*

Identification: Elongated, worm-like aquatic maggots without obvious antennae. Some species have dorsal and/or lateral projections. Some species are very elongated and thread like. The head is chitinized and obvious, projecting forward. Prolegs are absent or rarely present near the head. The breathing siphon at the tail is not obvious. **Size:** 1–10 mm. **Range:** Common and widespread. **Habitat:** Pools, ponds, wetlands, lakes, rivers, and streams. **Remarks:** The adults of these insects feed on the blood of other animals including insects and humans.

Biting Midges, No-See-Ums (Fig. 27s): Ceratopogonidae

Additional Remarks: An example of a specimen with lateral projections.

D. Blackflies (Fig. 27t): Simuliidae

Identification: Elongated, worm-like aquatic maggots with small, thin antennae and a bulbous tail end with an adhesive disc for attaching to rocks in swift current. The head is chitinized and obvious, projecting forward, and has two large "fans" projecting from each side of the mouth to filter food items from the water current. Prolegs are present near the head. **Size:** 1–8 mm. **Range:** Common and widespread. **Habitat:** Attached (sometimes in the millions) to rocks and other hard surfaces in riffles and waterfalls. Always in swift well-oxygenated water. **Remarks:** The adults of these insects feed on the blood of other animals including humans.

E. Soldierflies (Fig. 27u) Stratiomyidae:

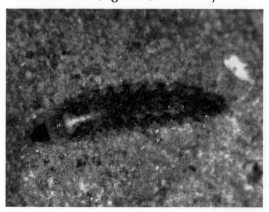

Identification: Broad, flattened, well-chitinized aquatic maggots without obvious antennae and strong projecting hairs or spines. Some species have an elongated breathing tube at the tail. The head is chitinized and obvious, projecting forward. Prolegs are absent. The breathing siphon at the tail is not obvious. **Size:** 1–20 mm. **Range:** Common and widespread. **Habitat:** Pools, ponds, wetlands, lakes, rivers, and streams. Some species are limited to hot springs or saline habitats in the west. **Remarks:** Some species are predators and others are herbivores.

F. Horseflies, Deerflies, Muleflies, Greenheads, Bulldogflies (Fig. 27v): Tabanidae, Pelecorhynchidae

Identification: Long, worm-like aquatic maggots without obvious antennae and bearing rings of creeping welts. The head is reduced, and often retracted into the body. Prolegs are absent. Often white in color, but may be patterned in yellow and black. **Size:** 1–60 mm. **Range:** Widespread and common, especially in natural areas. **Habitat:** Pools, ponds, wetlands, lakes, rivers, and streams. Burrowed in mud, sand, or silt. **Remarks:** The adults give painful bites. The larvae are active predators.

Horseflies, Deerflies, Muleflies, Greenheads, Bulldogflies (Fig. 27w): Tabanidae, Pelecorhynchidae

Additional Remarks: An example of a yellow and black specimen.

Watersnipe Flies (Fig. 27x): Athericidae

Identification: Long, worm-like aquatic maggots without obvious antennae, without creeping welts, but with several pairs of prolegs and a pair of tail filaments. The head is reduced and often retracted into the body. White, brown, or green in color. **Size:** 1–8 mm. **Range:** Widespread and common. **Habitat:** Occurs in riffles of rivers and streams. **Remarks:** The adult females bite. The larvae are predaceous.

G. Dagger Flies, Dance Flies, Balloon Flies (Fig. 27y): Empididae

Identification: Long, white aquatic maggots without obvious antennae or creeping welts. The head is reduced, and often retracted into the body. Many pairs of prolegs are present, and there are typically lobes at the tail. **Size:** 1–4 mm. **Range:** Widespread and common. **Habitat:** Flowing water; from swift riffles in rivers and streams to thin films of water moving over rocks in springs. **Remarks:** The larvae are predators.

H. Long-Legged Flies (Fig. 27z): Dolichopodidae

Identification: Long aquatic maggots without obvious antennae. Creeping welts present. The head is reduced, and often retracted into the body. Prolegs are absent, and the tail is lobed. White in color. **Size:** 1–4 mm. **Range:** Widespread and common. **Habitat:** Flowing water; from swift riffles in rivers and streams. **Remarks:** The larvae are predators.

I. Flower Flies (Fig. 27aa): Syrphidae

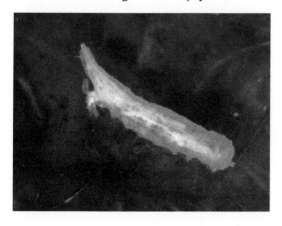

Identification: Oblong aquatic maggots without obvious antennae and a reduced and retracted head. Paired prolegs present, and the tail is elongated into a breathing siphon. Sometimes gills are present under the breathing siphon. White, yellow, or translucent in color. **Size:** 1–9 mm. **Range:** Widespread and common. **Habitat:** Found in water high in organic debris, such as stagnant wetlands and ponds, and in tree holes. **Remarks:** The larvae are voracious predators, often feeding on mosquito larvae. The adults are important pollinators.

J. Shore Flies, Brine Flies (Fig. 27ab): Ephydridae

Identification: Oblong, white, gray, or brown aquatic maggots without obvious antennae and a reduced and retracted head. Paired prologs or ventral creeping welts present, and the tail is elongated into a breathing siphon that is often split in two at the end. There is usually at least one pair of prologs at the end of the body. **Size:** 1–12 mm. **Range:** Widespread and common. **Habitat:** Many species live in stagnant waters, a few in flowing water, and many in saline pools and marshes. **Remarks:** The larvae are mostly filter feeders, but some are plant feeders and can be crop pests of rice, barley, and watercress.

K. Aquatic House Flies (Fig. 27ac): Muscidae: *Limnophora*

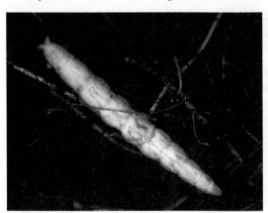

Identification: Oblong, white aquatic maggots with a greatly reduced and retracted head. Paired prologs on the last body segment the rest of the body with creeping welts. The tail is developed into a short, paired breathing siphon. **Size:** 1–10 mm. **Range:** Widespread but uncommon. **Habitat:** These animals live in aquatic mosses or similar vegetation. **Remarks:** Larval *Limnophora* are important predators on blackflies.

Glossary

asymmetrical: an individual animal having an irregular shape in the arrangement of body parts (as opposed to radial or bilateral symmetry); sponges are an example.

benthic (benthos): organisms living on or near the bottom of an aquatic habitat.

bilaterally symmetrical: having relatively equal distribution of body parts and shapes on both sides of a hypothetical plane bisecting the animal on a longitudinal axis; examples include flatworms and higher organisms, including humans.

biofilm: a thin organic and inorganic matrix coating a benthic surface with an embedded community of microinvertebrates, algae, fungi, and bacteria.

bivoltine: two generations per year.

cephalization: centralization of neural integrative structures (a brain) in an morphologically distinct head.

coelom: a fluid-filled cavity in the middle cell layer (mesoderm) of vertebrates and higher invertebrates such as annelid worms, snails, insects, crustaceans, etc.

colloidal organic matter: a mixture with properties between a fine suspension and dissolved solution; components of this colloid include proteins and amino acids.

concentric: in Mollusca, describing the growth lines of a snail operculum that lie entirely within each other (not forming a spiral); having a common center.

dioecious: having male and female reproductive organs in separate organisms.

ecdysis: shedding or molting of an older, smaller exoskeleton in arthropods and some other phyla.

encyst: the act of an organism developing a protective coating around itself which enables it to resist unfavorable environmental conditions (e.g., heat, cold, or lack or food or water); these are a diploid life stage rather than a gamete.

facultative: an optional process or relationship.

gemmule: an asexually produced reproductive structure in sponges which remains dormant through stressful periods, such as winter.

hermaphroditic: individual containing both male and female reproductive organs simultaneously or sequentially in life.

holoplankton: organisms living their entire lives in the plankton (e.g., copepods and water fleas).

iteroparous: reproducing repeatedly at multiple times during an organism's life, sometimes in the same field season; compare with semelparous.

lentic: standing water environments (e.g., lakes, ponds, wetlands, and both permanent and temporary pools).

lophophore: organized structure of ciliated tentacles used for capturing suspended food particles from the water, and probably also providing an important surface for gas exchange; occurring in ectoproct bryozoans and certain other phyla.

lotic: running water environments (e.g., creeks, rivers, and some springs).

medusa: in the phylum Cnidaria, a life history stage living primarily in the water column as plankton; a sexual stage of freshwater jellyfish that alternates with the polyp stage.

meroplankton: organisms living only part of their lives in the plankton (e.g., early instars of true midges).

multispiral: growth rings of an operculum of a snail comprised of many, slowly enlarging spirals.

multivoltine: having more than two broods or generations per year.

nauplius: earliest larval stage of many crustaceans.

nematocyst: complex stinging cells in hydra and jellyfish; used to subdue prey.

neuston: organisms living at or very near the air/water interface of a water body

ISBN 978-0-12-381426-5, DOI: 10.1016/B978-0-12-381426-5.00028-4

ocellus (ocelli): simple eye consisting of a single, bead-like lens, occurring singly or in small groups; distinct from a compound eye.

oviparous: females that produce eggs that undergo embryonic development and hatch outside the body of the female; see viviparous and ovoviviparous.

ovoviviparous: females which produce eggs that undergo embryonic development inside the female and that hatch just before or soon after being released; see oviparous and viviparous.

parthenogenesis: sexual reproduction in which egg development occurs without fertilization by a male.

paucispiral: growth lines of a snail operculum, present as a few, rapidly enlarging spirals.

periostracum: proteinaceous outer layer of mollusc shells secreted by the mantle edge.

periphyton: microalgae growing in a thin layer on mostly vascular plants; sometimes used loosely for microalgae growing on other substances.

pharynx: portion of the alimentary tract; in flatworms this is muscular and can be extended outward from the body to collect prey.

plankton: bacteria, protozoa, algae, and microscopic animals like copepods, rotifers, and water fleas that live all or part of their lives in the water column of aquatic ecosystems and which are at the mercy of water currents.

planospiral: having shell whorls confined to a single plane; coiled but without an elevated spire.

polyp: in the phylum Cnidaria, a life history stage living on the bottom with tentacles projecting away from the point of attachment; may be a sexual and/or asexual stage.

polypide: the partially protrusible organ system in bryozoans containing the lophophore and other internal organ systems and which is surrounded by a body wall.

radially symmetrical: having body parts arranged around a central point; an animal body structure that tends to be circular in shape, such as in a freshwater jellyfish or hydra.

radula: a rasping, file-like structure in the foot of a snail that is used to scrap food from a surface.

semelparous: breeding only once during an organism's life; one massive reproductive effort; compare with iteroparous.

semivoltine: requiring more than 1 year to produce a single generation.

statocyst: gravity-orientation organ containing mineral inclusions in a vesicle lined with sensory cilia.

symbiont: one member of a symbiotic relationship.

troglobite: obligate cave-dweller, often characterized in crustaceans by a reduction in eye structure and a lack of pigments; also, frequently demonstrating K-strategies (e.g., delayed reproduction, long lives).

troglophile: organism living in a cave in a nonobligate relationship; such organisms usually complete their life history in the cave and commonly have conspecifics surviving outside the cave; contrasts with troglobite.

univoltine: having one brood or generation per year.

verge: penis, or organ that bears the penis, in snails.

vernal pool: an ephemeral or temporary pool of water that forms in the spring and dries in the summer.

viviparous: type of reproduction in which the young develop internally and are maintained and nourished by the mother before birth; see oviparous and ovoviviparous.

zooids: one of the physically connected, asexually replicated, morphologic units that comprise a colony.

References and Suggested Readings

Adler, P.H., Currie, D.C., Wood, D.M., 2004. The Black Flies (Simuliidae) of North America. Cornell University Press, Ithaca, NY.

Allan, J.D., Castillo, M.M., 2007. Stream Ecology: Structure and Function of Running Waters, second ed. Springer, New York.

Bliss, D.E. (editor-in-chief), 1982–1985. The Biology of Crustacea, vols. 1–10, Academic Press, New York.

Bowles, E.E., Aziz, K., Knight, C.L., 2000. *Macrobrachium* (Decapoda: Caridea: Palaemonidae) in the contiguous United States: a review of the species and an assessment of threats to their survival. Journal of Crustacean Biology 29, 158–171.

Brendonck, L., Rogers, D.C., Olesen, J., Weeks, S., Hoeh, W.R., 2008. Global diversity of large branchiopods (Crustacea: Branchiopoda) in freshwater. Hydrobiologia 595, 167–176.

Brown, K.M., Lang, B., Perez, K.E., 2008. The conservation ecology of North American pleurocerid and hydrobiid gastropods. Journal of the North American Benthological Society 27, 484–495.

Brown, K.M., Lydeard, C., 2010. Mollusca: Gastropoda. In: Thorp, J.H., Covich, A.P. (Eds.), Ecology and Classification of North American Freshwater Invertebrates. Academic Press, Elsevier, London, UK, Chapter 10, pp. 277–307.

Burch, J.B., 1989. North American Freshwater Snails. Malacological Publications, Hamburg, MI, 365 pp.

Canada, A., Branch, R. 1981. Manual of Nearctic Diptera, vol. 1. Monograph No. 27. Canadian Government Publishing Centre, Supply and Services Canada, Hull, Que., Canada, 674 p.

Canada, A., Branch, R. 1987. Manual of Nearctic Diptera, vol. 2. Monograph No. 28. Canadian Government Publishing Centre, Supply and Services Canada, Hull, Que., Canada, 658 p.

Cannon, L.R.G., 1986. Turbellaria of the World. A Guide to Families and Genera. Queensland Museum, Brisbane.

Carico, J.E., 1973. The nearctic species of the genus *Dolomedes* (Araneae: Pisauridae). Bulletin of the Museum of Comparative Zoology 144, 435–488.

Cook, D.R., 1974. Water mite genera and subgenera. Memoirs of the American Entomological Institute 21, 860.

Covich, A.P., Thorp, J.H., Rogers, D.C., 2010. Introduction to the subphylum Crustacea. In: Thorp, J.H., Covich, A.P. (Eds.), Ecology and Classification of North American Freshwater Invertebrates. Academic Press, Elsevier, London, UK, Chapter 18, pp. 695–723.

Culver, D.C., White, W.B. (Eds.), 2004. Encyclopedia of Caves. Academic Press, Elsevier, San Diego, CA, 654 p.

Cummings, K.S., Graf, D.L., 2010. Mollusca: Bivalvia. In: Thorp, J.H., Covich, A.P. (Eds.), Ecology and Classification of North American Freshwater Invertebrates. Academic Press, Elsevier, London, UK, Chapter 11, pp. 309–384.

DeWalt, R.E., Resh, V.H., Hilsenhoff, W.L., 2010. Diversity and classification of insects and collembola. In: Thorp, J.H., Covich, A.P. (Eds.), Ecology and Classification of North American Freshwater Invertebrates. Academic Press, Elsevier, London, UK, Chapter 16, pp. 587–657.

Devries, D.R., 1992. The fresh-water jellyfish *Craspedacusta sowerbyi* – a summary of its life-history, ecology, and distribution. Journal of Freshwater Ecology 7, 7–16.

© 2011 Elsevier Inc. All rights reserved.
ISBN 978-0-12-381426-5, DOI: 10.1016/B978-0-12-381426-5.00029-6

Dillon, R.T., 2000. The Ecology of Freshwater Molluscs. Cambridge University Press, New York, NY. 509 p.

Dodds, W.K., 2002. Freshwater ecology: Concepts and Environmental Applications. Academic Press, Elsevier, San Diego, CA, 569 p.

Dodson, S.I., Cáceres, C.E., Rogers, D.C., 2010. Cladocera and other Branchiopoda. In: Thorp, J.H., Covich, A.P. (Eds.), Ecology and Classification of North American Freshwater Invertebrates. Academic Press, Elsevier, London, UK, Chapter 20, pp. 774–827.

Dodson, S.I., Cooper, S.D., 1983. Trophic relationships of the freshwater jellyfish *Craspedacusta* sowerbii Lankester 1880. Limnology and Oceanography 28, 345–351.

Dondale, C.D., Redner, J.H., 1990. The wolf spiders, nurseryweb spiders, and lynx spiders of Canada and Alaska (Araneae: Lycosidae, Pisauridae, and Oxyopidae). The Insects and Arachnids of Canada, Part 17, Agriculture Canada, Ottawa, Canada.

Frost, T.M., de Nagy, G.S., Gilbert, J.J., 1982. Population dynamics and standing biomass of the freshwater sponge, *Spongilla lacustris*. Ecology 63, 1203–1210.

Govedich, F.R., Bain, B.A., Moser, W.E., Gelder, S.R., Davies, R.W., Brinkhurst, R.O., 2010. Annelida (Clitellata): Oligochaeta, Branchiobdellida, Hirudinida, and Acanthobdellida. In: Thorp, J.H., Covich, A.P. (Eds.), Ecology and Classification of North American Freshwater Invertebrates. Academic Press, Elsevier, London, UK, Chapter 12, pp. 385–436.

Graf, D.L., Cummings, K.S., 2007. Review of the systematics and global diversity of freshwater mussel species (Bivalvia: Unionoida). Journal of Molluscan Studies 73, 291–314.

Hanelt, B., Janovy, J. Jr., 2004. Life cycle and paratenesis of American gordiids (Nematomorpha: Gordiida). Journal of Parasitology 90, 240–244.

Hershey, A.E., Lamberti, G.A., Chaloner, D.T., Northington, R.M., 2010. Aquatic insect ecology. In: Thorp, J.H., Covich, A.P. (Eds.), Ecology and Classification of North American Freshwater Invertebrates. Academic Press, Elsevier, London, UK, Chapter 17, pp. 659–694.

Hobbs, H.H.I.I.I., Lodge, D.M., 2010. Decapoda. In: Thorp, J.H., Covich, A.P. (Eds.), Ecology and Classification of North American Freshwater Invertebrates. Academic Press, Elsevier, London, UK, Chapter 22, pp. 901–967.

Holdich, D.M. (Ed.), 2002. Biology of Freshwater Crayfish. Blackwell Science, Oxford, 702 p.

Hutchinson, G.E., 1967. A Treatise on Limnology. II. Introduction to Lake Biology and Limnoplankton. Wiley, New York, 1115 p.

Hutchinson, G.E., 1993. A Treatise on Limnology. IV. The Zoobenthos. Wiley, New York, 944 p.

Kenk, R., 1972. EPA Biota of Freshwater Ecosystems, Identification Manual, vol. 1, Freshwater planarians (Turbellaria) of North America. Water Pollution Control Research Series, U.S. Environmental Protection Agency, Washington, D.C. pp. 1–81.

King, D.K., King, R.H., Miller, A.C., 1988. Morphology and ecology of *Urnatella gracilis* Leidy, (Entoprocta), a freshwater macroinvertebrate from artificial riffles of the Tombigbee River, Mississippi. Journal of Freshwater Ecology 4, 351–359.

Kolasa, J., Tyler, S., 2010. Flatworms: turbellarians and Nemertea. In: Thorp, J.H., Covich, A.P. (Eds.), Ecology and Classification of North American Freshwater Invertebrates. Academic Press, Elsevier, London, UK, Chapter 6, pp. 143–161.

Krantz, G.W., Walter, D.E., 2009. A Manual of Acarology, , third ed. Texas Technolgy University Press, Lubbock, Texas, 816 p.

Lampert, W., Sommer, U., 2007. Limnoecology, The Ecology of Lakes and Streams, second ed. Oxford University, Oxford, UK, 324 p.

Lee, T., 2004. Morphology and phylogenetic relationships of genera of North American Sphaeriidae (Bivalvia, Veneroida). American Malacological Bulletin 19, 1–13.

Lenhoff, H.M., 1983. Hydra: Research Methods. Plenum Press, New York, 463 p.

Likens, G.E. (editor-in-chief), 2009. Encyclopedia of Inland Waters. Academic Press, San Diego, CA, 6492 p.

Mackie, G.L., 2007. Biology of freshwater corbiculid and sphaeriid clams of North America. Ohio Biological Survey Bulletin (New Series) 15, 436.

McLaughlin, P.A., Camp, D.K., Angel, M.V., Bousfield, E.L., Brunnel, P., Brusca, R.C., et al., 2005. Common and Scientific Names of Aquatic Invertebrates from the United States and Canada: Crustaceans. American Fisheries Society, Special Publication 31, Bethesda, MD.

Merritt, R.W., Cummins, K.W., Berg, M.B., 2008. Introduction to the Aquatic Insects of North America. Kendall/ Hunt, Dubuque, Iowa, 1214 p.

Nalepa, T.F., Schloesser, D.W. (Eds.), 1993. Zebra Mussels: Biology, Impacts, and Control. Lewis Publishers, Boca Raton, FL.

Needham, J.G., Westfall, M.J. Jr., May, M.L., 2000. Dragonflies of North America, revised ed. Scientific Publishers, Gainesville, FL.

Peck, S.B., 1998. A summary of diversity and distribution of the obligate cave-inhabiting fauna of the United States and Canada. Journal of Cave and Karst Studies 60, 18–26.

Poinar, Jr.,G., 2008. Global diversity of hairworms (Nematomorpha: Gordiaceae) in freshwater. In: Balian, E.V., Lévêque, C., Segers, H., Martens, K. (Eds.), Freshwater Animal Diversity Assessment. Hybrobiologia, vol. 198, Springer, Netherland, pp. 79–83.

Poinar, G.O. Jr., 2010. Nematoda and Nematomorpha. In: Thorp, J.H., Covich, A.P. (Eds.), Ecology and Classification of North American Freshwater Invertebrates. Academic Press, Elsevier, London, UK, Chapter 9, pp. 237–276.

Poly, W.J., 2008. Global diversity of fishlice (Crustacea: Branchiura: Argulidae). Hydrobiologia 595, 209–212.

Proctor, H.C., Pritchard, G., 1989. Neglected predators: water mites (Acari: Parasitengona: Hydrachnellae) in freshwater communities. Journal of the North American Benthological Society 8, 100–111.

Reid, J.W., Williamson, C.E., 2010. Copepoda. In: Thorp, J.H., Covich, A.P. (Eds.), Ecology and Classification of North American Freshwater Invertebrates. Academic Press, Elsevier, London, UK, Chapter 21, pp. 829–899.

Reiswig, H.M., Frost, T.M., Ricciardi, A., 2010. Porifera. In: Thorp, J.H., Covich, A.P. (Eds.), Ecology and Classification of North American Freshwater Invertebrates Academic Press, London, UK, Chapter 4, pp. 91–123.

Resh, V.H., Cardé, R.T. (Eds.), 2009. Encyclopedia of Insects, second ed. Elsevier, San Diego, CA, 1132 p.

Ricciardi, A., Reiswig, H.M., 1993. Freshwater sponges (Porifera, Spongillidae) of eastern Canada: taxonomy, distribution, and ecology. Canadian Journal of Zoology 71, 665–682.

Ricciardi, A., Reiswig, H., 1994. Taxonomy, distribution, and ecology of the freshwater bryozoans (Ectoprocta) of eastern Canada. Canadian Journal of Zoology 72, 339–359.

Rogers, D.C., 2009. Branchiopoda (Anostraca, Notostraca, Laevicaudata, Spinicaudata, Cyclestherida). In: Likens, G.E. (Ed.), Encyclopedia of Inland Waters. Academic Press, Elsevier, London, UK, pp. 242–249.

Schram, F.R., 1986. Crustacea. Oxford University Press, New York, 606 pp.

Slobodkin, L.B., Bossert, P.E., 2010. Cnidaria. In: Thorp, J.H., Covich, A.P. (Eds.), Ecology and Classification of North American Freshwater Invertebrates. Academic Press, Elsevier, London, UK, Chapter 5, pp. 125–142.

Smith, B.P., 1988. Host-parasite interaction and impact of larval water mites on insects. Annual Review of Entomology 33, 487–507.

Smith, I.M., Cook, D.R., Smith, B.P., 2010. Water mites (Hydrachnidiae) and other arachnids. In: Thorp, J.H., Covich, A.P. (Eds.), Ecology and Classification of North American Freshwater Invertebrates. Academic Press, Elsevier, London, UK, Chapter 15, pp. 485–586.

Smith, A.J., Delorme, L.D., 2010. Ostracoda. In: Thorp, J.H., Covich, A.P. (Eds.), Ecology and Classification of North American Freshwater Invertebrates. Academic Press, Elsevier, London, UK, Chapter 19, pp. 725–771.

Stewart, K.W., Stark, B.P., 2002. Nymphs of North American Stonefly Genera (Plecoptera), second ed. The Caddis Press, Columbus, OH, 510 pp.

Sturm, C.F., Pearce, T.A., Valdes, A., 2006. The Mollusks: A Guide to their Study, Collection, and Preservation. American Malacological Society, Pittsburgh, PA, USA, Universal Publishers, Boca Raton, FL.

Thorp, J.H., 2009. Arthropoda and related groups. In: Resh, V.H., Cardé, R.T. (Eds.), Encyclopedia of Insects. Academic Press, San Diego, CA, pp. 50–56.

Thorp, J.H., Covich, A.P., 2010. Ecology and Classification of North American Freshwater Invertebrates. Academic Press, Elsevier, London, UK, 1021 p.

Thorp, J.H., Thoms, M.C., Delong, M.D., 2008. The Riverine Ecosystem Synthesis. Academic Press, Elsevier, Boston, MA, 208 p.

Tyler, S., Schilling, S., Hooge, M., Bush, L.F. (comp.) 2008. Turbellarian taxonomic database. Version 1.5; http://turbellaria.umaine.edu.

Walter, D.E., Lindquist, E.E., Smith, I.M., Cook, D.R., Krantz, G.W. (In press). Sphaerolichida and Prostigmata. In: Krantz, G.W., Walter, D.E. (Eds.), A Manual of Acarology, third ed. Texas Technology University Press, Lubbock, TX, Chapter 13, 816 p.

Westfall, M.J., May, M.L., Jr., 1996. Damselflies of North America. Scientific Publishers, Gainesville, FL.

Wetzel, R.G., 2001. Limnology, third ed. Academic Press, San Diego, CA, 1066 p.

Wiggins, G.B., 1996. Larvae of the North American Caddisfly Genera (Trichoptera), second ed. University of Toronto Press, Toronto.

Wiggins, G.B., Mackay, R.J., Smith, I.M., 1980. Evolutionary and ecological strategies of animals in annual temporary pools. Archive fur Hydrobiologie/Suppl. 58 (1), 97–206.

Wilson, G.D.F., 2008. Global diversity of isopod crustaceans (Crustacea; Isopoda) in freshwater. Hydrobiologia 595, 231–240.

Wood, T.S., 2010. Bryozoans. In: Thorp, J.H., Covich, A.P. (Eds.), Ecology and Classification of North American Freshwater Invertebrates. Academic Press, Elsevier, London, UK, Chapter 13, pp. 437–454.

Yeo, D.C., Ng, P.K.L., Cumberlidge, N., Magalhães, C., Daniels, S.R., Campos, M.R., 2008. Global diversity of crabs (Crustacea: Decapoda: Brachyura) in fresh water. Hydrobiologia 595, 275–286.

Index

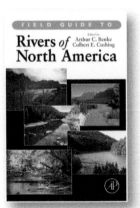

Printed and bound by CPI Group (UK) Ltd, Croydon, CR0 4YY

03/10/2024

01040399-0007